Weil-Petersson Metric on the Universal Teichmüller Space

of the
American Mathematical Society

Number 861

Weil-Petersson Metric on the Universal Teichmüller Space

Leon A. Takhtajan
Lee-Peng Teo

September 2006 • Volume 183 • Number 861 (first of 4 numbers) • ISSN 0065-9266

American Mathematical Society
Providence, Rhode Island

2000 *Mathematics Subject Classification.* Primary 30F60; Secondary 30C55, 32G15, 46E20, 58B20, 58B25.

Library of Congress Cataloging-in-Publication Data

Takhtajan, Leon A.
  Weil-Petersson metric on the universal Teichmüller space / Leon A. Takhtajan.
    p. cm. — (Memoirs of the American Mathematical Society, ISSN 0065-9266 ; no. 861)
  "Volume 183, number 861 (first of 4 numbers)."
  Includes bibliographical references.
  ISBN-13: 978-0-8218-3936-2 (alk. paper)
    1. Teichmüller spaces. 2. Univalent functions. 3. Riemann surfaces. I. Teo, Lee-Peng, 1975– II. Title. III. Series.

QA3.A57   no. 861
[QA337]
510 s—dc22
[515′.94]
                                                                                            2006045733

Memoirs of the American Mathematical Society

This journal is devoted entirely to research in pure and applied mathematics.

Subscription information. The 2006 subscription begins with volume 179 and consists of six mailings, each containing one or more numbers. Subscription prices for 2006 are US$624 list, US$499 institutional member. A late charge of 10% of the subscription price will be imposed on orders received from nonmembers after January 1 of the subscription year. Subscribers outside the United States and India must pay a postage surcharge of US$31; subscribers in India must pay a postage surcharge of US$43. Expedited delivery to destinations in North America US$35; elsewhere US$130. Each number may be ordered separately; *please specify number* when ordering an individual number. For prices and titles of recently released numbers, see the New Publications sections of the *Notices of the American Mathematical Society*.

Back number information. For back issues see the *AMS Catalog of Publications*.

Subscriptions and orders should be addressed to the American Mathematical Society, P. O. Box 845904, Boston, MA 02284-5904, USA. *All orders must be accompanied by payment.* Other correspondence should be addressed to 201 Charles Street, Providence, RI 02904-2294, USA.

Copying and reprinting. Individual readers of this publication, and nonprofit libraries acting for them, are permitted to make fair use of the material, such as to copy a chapter for use in teaching or research. Permission is granted to quote brief passages from this publication in reviews, provided the customary acknowledgment of the source is given.

Republication, systematic copying, or multiple reproduction of any material in this publication is permitted only under license from the American Mathematical Society. Requests for such permission should be addressed to the Acquisitions Department, American Mathematical Society, 201 Charles Street, Providence, Rhode Island 02904-2294, USA. Requests can also be made by e-mail to reprint-permission@ams.org.

Memoirs of the American Mathematical Society is published bimonthly (each volume consisting usually of more than one number) by the American Mathematical Society at 201 Charles Street, Providence, RI 02904-2294, USA. Periodicals postage paid at Providence, RI. Postmaster: Send address changes to Memoirs, American Mathematical Society, 201 Charles Street, Providence, RI 02904-2294, USA.

© 2006 by the American Mathematical Society. All rights reserved.
Copyright of this publication reverts to the public domain 28 years
after publication. Contact the AMS for copyright status.
This publication is indexed in *Science Citation Index*®, *SciSearch*®, *Research Alert*®, *CompuMath Citation Index*®, *Current Contents*®/*Physical, Chemical & Earth Sciences*.
Printed in the United States of America.

∞ The paper used in this book is acid-free and falls within the guidelines
established to ensure permanence and durability.
Visit the AMS home page at http://www.ams.org/

10 9 8 7 6 5 4 3 2 1   11 10 09 08 07 06

Contents

Introduction   1

Chapter 1. Curvature Properties and Chern Forms   7
  1. The universal Teichmüller space   7
  1.1. Teichmüller theory   7
  1.1.1. Main definitions   7
  1.1.2. The group structure   8
  1.1.3. The Bers embedding   9
  1.1.4. The complex structure   9
  1.1.5. The universal Teichmüller curve   10
  1.2. Homogeneous spaces of $\mathrm{Homeo}_{qs}(S^1)$ 11
  1.2.1. Conformal welding   11
  1.2.2. The horizontal and vertical subspaces   14
  1.2.3. The isomorphisms of the tangent spaces   15
  1.3. Teichmüller spaces and Teichmüller curves of Fuchsian groups   17
  1.4. Resolvent kernel   18
  1.5. Variational formulas   19
  2. $T(1)$ as a Hilbert manifold   22
  2.1. Hilbert space structure on tangent spaces   22
  2.2. The L^2-estimates   27
  2.3. The Hilbert manifold structure of $T(1)$ 29
  2.4. Integral manifolds of the distribution $\mathfrak{D}_T$ 31
  3. $T_0(1)$ as a topological group   32
  4. Velling-Kirillov and Weil-Petersson metrics   36
  4.1. Velling-Kirillov metric on the universal Teichmüller curve   36
  4.2. Weil-Petersson metric on the universal Teichmüller space   37
  5. Characteristic forms of the universal Teichmüller curve   38
  5.1. The form $c_1(V)$ as Velling-Kirillov symplectic form   39
  5.2. The Chern form $c_1(V)$ and the resolvent kernel   41
  5.3. Mumford-Morita-Miller characteristic forms   43
  6. First and second variations of the hyperbolic metric   44
  6.1. The first variation   44
  6.2. The second variation   44
  7. Riemann curvature tensor   46
  7.1. The first variation of the Weil-Petersson metric   46
  7.2. The second variation of the Weil-Petersson metric   50
  7.3. Ricci and sectional curvatures   54
  8. Finite-dimensional Teichmüller spaces   56

Chapter 2. Kähler Potential and Period Mapping   65

1.	Hilbert spaces and univalent functions	65
2.	Grunsky operators for $\mathcal{T}_0(1)$	70
2.1.	Grunsky coefficients and operators	70
2.2.	Fredholm eigenvalues and Fredholm determinant	78
2.3.	Period matrix of 1-forms	82
3.	Variations of the functions S_1 and S_2	83
3.1.	The first variation of S_2	84
3.2.	The first variation of S_1	85
4.	Weil-Petersson potential	95
4.1.	Weil-Petersson potential on $\mathcal{T}_0(1)$	95
4.2.	Weil-Petersson potential on $T(1)$	97
5.	The period mapping	97
5.1.	KYNS period mapping	98
5.2.	Embeddings into the Segal-Wilson universal Grassmannian	101

Appendix A.	The Hilbert Manifold Structure of $\mathcal{T}_0(1)$	105
Appendix B.	The Period Mapping $\hat{\mathscr{P}}$	109
Bibliography		117

Abstract

In this memoir, we prove that the universal Teichmüller space $T(1)$ carries a new structure of a complex Hilbert manifold and show that the connected component of the identity of $T(1)$ — the Hilbert submanifold $T_0(1)$ — is a topological group. We define a Weil-Petersson metric on $T(1)$ by Hilbert space inner products on tangent spaces, compute its Riemann curvature tensor, and show that $T(1)$ is a Kähler-Einstein manifold with negative Ricci and sectional curvatures. We introduce and compute Mumford-Miller-Morita characteristic forms for the vertical tangent bundle of the universal Teichmüller curve fibration over the universal Teichmüller space. As an application, we derive Wolpert curvature formulas for the finite-dimensional Teichmüller spaces from the formulas for the universal Teichmüller space. We study in detail the Hilbert manifold structure on $T_0(1)$ and characterize points on $T_0(1)$ in terms of Bers and pre-Bers embeddings by proving that the Grunsky operators B_1 and B_4, associated with the points in $T_0(1)$ via conformal welding, are Hilbert-Schmidt. We define a "universal Liouville action" — a real-valued function S_1 on $T_0(1)$, and prove that it is a Kähler potential of the Weil-Petersson metric on $T_0(1)$. We also prove that S_1 is $-\frac{1}{12\pi}$ times the logarithm of the Fredholm determinant of associated quasi-circle, which generalizes classical results of Schiffer and Hawley. We define the universal period mapping $\hat{\mathscr{P}} : T(1) \to \mathscr{B}(\ell^2)$ of $T(1)$ into the Banach space of bounded operators on the Hilbert space ℓ^2, prove that $\hat{\mathscr{P}}$ is a holomorphic mapping of Banach manifolds, and show that $\hat{\mathscr{P}}$ coincides with the period mapping introduced by Kurillov and Yuriev and Nag and Sullivan. We prove that the restriction of $\hat{\mathscr{P}}$ to $T_0(1)$ is an inclusion of $T_0(1)$ into the Segal-Wilson universal Grassmannian, which is a holomorphic mapping of Hilbert manifolds. We also prove that the image of the topological group S of symmetric homeomorphisms of S^1 under the mapping $\hat{\mathscr{P}}$ consists of compact operators on ℓ^2.

The results of this memoir were presented in our e-prints: *Weil-Petersson metric on the universal Teichmuller space I. Curvature properties and Chern forms*, arXiv:math.CV/0312172 (2003), and *Weil-Petersson metric on the universal Teichmuller space II. Kahler potential and period mapping*, arXiv:math.CV/0406408 (2004).

Received by the editor January 26, 2005.

2000 *Mathematics Subject Classification.* 30F60 (Primary) 30C55, 32G15, 46E20, 58B20, 58B25 (Secondary).

Key words and phrases. Universal Teichmüller space, Bers coordinates, Bers embedding, Hilbert manifold structure, Weil-Petersson metric, variation of the hyperbolic metric, Riemann curvature tensor, Grunsky operators, Kähler potential, Fredholm determinant, universal Liouville action, period mapping.

Introduction

The universal Teichmüller space $T(1)$ is the simplest Teichmüller space that bridges spaces of univalent functions and general Teichmüller spaces. Introduced by Bers [**Ber65, Ber72, Ber73**], the universal Teichmüller space is an infinite-dimensional complex manifold modeled on a Banach space. It contains Teichmüller spaces of Riemann surfaces as complex submanifolds. The universal Teichmüller space $T(1)$ also came to the forefront with the advent of string theory. Its complex submanifold — an infinite-dimensional complex Fréchet manifold $\text{Möb}(S^1)\backslash \text{Diff}_+(S^1)$, plays an important role in one of the approaches to non-perturbative bosonic closed string field theory based on Kähler geometry [**BR87a, BR87b**]. The manifold $\text{Möb}(S^1)\backslash \text{Diff}_+(S^1)$ — a homogeneous space of the Lie group $\text{Diff}_+(S^1)$, also has an interpretation as a coadjoint orbit of the Bott-Virasoro group, and as such carries a natural right-invariant Kähler metric [**Kir87, KY87**].

The complex geometry of the finite-dimensional Teichmüller spaces — Teichmüller spaces $T(\Gamma)$ of cofinite Fuchsian groups, has been extensively studied in the context of Ahlfors-Bers deformation theory of complex structures on Riemann surfaces. In particular, A. Weil defined a natural Hermitian metric on $T(\Gamma)$ by the Petersson inner product on the tangent spaces. Called Weil-Petersson metric, it was shown to be a Kähler metric by Weil and Ahlfors. In his seminal paper [**Ahl62**] Ahlfors has studied the curvature properties of the Weil-Petersson metric. In particular, he proved that the Bers coordinates on $T(\Gamma)$ are geodesic at the origin, and computed the Riemann curvature tensor of the Weil-Petersson metric in terms of multiple principal value integrals. Using these formulas, Ahlfors proved that $T(\Gamma)$ has negative Ricci, holomorphic sectional, and scalar curvatures. Further results have been obtained by Royden [**Roy75**]. Wolpert re-examined Ahlfors' approach in [**Wol86**]. He developed a different method for computing Riemann and Ricci curvature tensors, and obtained explicit formulas in terms of the resolvent kernel of the Laplace operator of the hyperbolic metric on the corresponding Riemann surface.

Curvature properties of the infinite-dimensional manifold $\text{Möb}(S^1)\backslash \text{Diff}_+(S^1)$ have been studied by Kirillov and Yuriev [**KY87**], and by Bowick and Rajeev [**BR87a, BR87b**]. In particular, they computed the Riemann curvature tensor of the right-invariant Kähler metric and proved that $\text{Möb}(S^1)\backslash \text{Diff}_+(S^1)$ is a Kähler-Einstein manifold.

Since both the finite-dimensional Teichmüller spaces $T(\Gamma)$ and the homogeneous space $\text{Möb}(S^1)\backslash \text{Diff}_+(S^1)$ are complex submanifolds of $T(1)$, it is natural to investigate whether the latter space carries a "universal" Kähler metric which can be pulled back to the submanifolds. The immediate difficulty is that the universal Teichmüller space $T(1)$ is a complex Banach manifold, so that its tangent spaces do not carry Hermitian metric. Nag and Verjovsky [**NV90**] were the first to address this problem. They have shown that the Kähler metric on $\text{Möb}(S^1)\backslash \text{Diff}_+(S^1)$

is the pull-back of a certain Hermitian metric defined on a Hilbert subspace of the tangent space at the origin of $T(1)$. The latter metric is analogous to the Weil-Petersson metric on finite-dimensional Teichmüller spaces. However, finite-dimensional Teichmüller spaces $T(\Gamma)$ embed into $T(1)$ transversally to the Hilbert subspace, so that the Weil-Petersson metric on $T(\Gamma)$ can not be pulled back from $T(1)$. Nevertheless, following a suggestion by Velling, Nag and Verjovsky [**NV90**] have shown that the Weil-Petersson metric on $T(\Gamma)$ can be obtained by a certain "averaging" procedure using Patterson's uniform distribution of the "lattice points" of a cofinite Fuchsian group Γ in the hyperbolic plane. The major open problem is to define the Weil-Petersson metric on the whole space $T(1)$, to study its curvature properties, and to find relation between curvatures of this metric and the Weil-Petersson metric on finite-dimensional Teichmüller spaces[1].

In this memoir, we introduce Weil-Petersson metric on the universal Teichmüller space $T(1)$ and study its curvature properties. In Chapter 1 we prove that $T(1)$ carries a new structure of a Hilbert manifold such that in the underlying topology $T(1)$ has uncountably many components. We prove that the connected component of the identity in $T(1)$, the Hilbert submanifold $T_0(1)$, is a topological group and describe its image under the Bers embedding $\beta : T(1) \to A_\infty(\mathbb{D})$. We define the Weil-Petersson metric on $T(1)$ by Hilbert space inner products on tangent spaces. We re-examine the Ahlfors original computation [**Ahl62**] of the second variation of the hyperbolic metric and of the Riemann tensor for the finite-dimensional Teichmüller spaces in terms of the principal value integrals. We show how to extend the Ahlfors' method to the case of the universal Teichmüller space and how to convert formulas using principal value integrals into closed expressions using resolvent kernel of the Laplace operator on the hyperbolic plane. Our results extend the Wolpert's formulas [**Wol86**] to the infinite-dimensional Hilbert manifold $T(1)$. We also prove that $T(1)$ is a Kähler-Einstein manifold with negative Ricci and sectional curvatures. Using the averaging procedure, we derive Wolpert's curvature formulas [**Wol86**] for the finite-dimensional Teichmüller spaces from the curvature formulas for the universal Teichmüller space. We introduce and compute Mumford-Morita-Miller characteristic forms for the vertical tangent bundle associated with the fibration $\pi : \mathcal{T}(1) \to T(1)$, where $\mathcal{T}(1)$ is the universal Teichmüller curve. Here again we consider $T(1)$ and $\mathcal{T}(1)$ as Hilbert manifolds and show that the integration over the fibers operation, used in the definition of Mumford-Morita-Miller characteristic forms, is well-defined.

In Chapter 2 we study in detail the Hilbert manifold structure of $T(1)$ and establish relations between the Hilbert submanifold $T_0(1)$ — the connected component of the identity in $T(1)$, and classical Grunsky operators B_l, $l = 1, 2, 3, 4$, associated with the conformal welding. We characterize $T_0(1)$ in terms of the pre-Bers embedding $\hat{\beta} : T(1) \to A_\infty^1(\mathbb{D})$ and prove that the Grunsky operators B_1 and B_4 associated with the points in $T_0(1)$ are Hilbert-Schmidt. We establish the relation between eigenvalues of Grunsky operators and classical Fredholm eigenvalues, generalizing Schiffer's result for C^3 curves [**Sch81**]. We prove that the logarithm of the Fredholm determinant of the operator $I - B_1 B_1^*$ associated with points in $T_0(1)$ (or, which is the same, of the Fredholm determinant of $I - B_4 B_4^*$) is, up to a constant, a Kähler potential for the Weil-Petersson metric on $T_0(1)$. We prove the

---

[1] See the remark on p. 136 in [**NV90**].

explicit formula for this Fredholm determinant, expressing it as the "universal Liouville action". Using Grunsky operators, we define the universal period mapping $\mathscr{P}$ of $T_0(1)$ into the Hilbert space $\mathscr{S}_2$ of Hilbert-Scmidt operators on the Hilbert space ℓ^2, as well as the mapping $\hat{\mathscr{P}}$ of $T(1)$ into the Banach space $\mathscr{B}(\ell^2)$ of bounded operators on ℓ^2. We prove that $\mathscr{P}$ and $\hat{\mathscr{P}}$ are holomorphic mappings of Hilbert and Banach manifolds respectively. We show that the mapping $\hat{\mathscr{P}}$ coincides with the period mapping, first introduced by Kirillov and Yuriev [**KY88**] for the homogenous space $\mathrm{M\ddot{o}b}(S^1)\backslash\mathrm{Diff}_+(S^1)$, studied in detail by Nag [**Nag92**], and then extended to $T(1)$ by Nag and Sullivan [**NS95**][2]. Finally, we prove that the image of the topological group S of symmetric homeomorphisms of S^1 under the period mapping $\hat{\mathscr{P}}$ is $\mathscr{S}_\infty \cap \hat{\mathscr{P}}(T(1))$, where $\mathscr{S}_\infty$ is the ideal of the Banach algebra $\mathscr{B}(\ell^2)$ consisting of compact operators on ℓ^2.

Here is the more detailed content of the paper, starting with Chapter 1. In Section 1 we present necessary facts from Teichmüller theory, mainly following classical monographs by Ahlfors [**Ahl87**], Lehto [**Leh87**] and Nag [**Nag88**]. Namely, in Section 1.1 we briefly cover: the main definitions, the group structure of the universal Teichmüller space $T(1)$, the Bers embedding, structure of $T(1)$ as an infinite-dimensional complex Banach manifold modeled on the complex Banach space $A_\infty(\mathbb{D})$, and the basic properties of the universal Teichmüller curve $\pi : \mathcal{T}(1) \to T(1)$. In Section 1.2 we realize $T(1)$ and $\mathcal{T}(1)$ as homogeneous spaces of the group $\mathrm{Homeo}_{qs}(S^1)$ of quasi-symmetric homeomorphisms of S^1, and by using conformal welding we identify $T(1)$ and $\mathcal{T}(1)$ with the spaces of univalent functions on the unit disk $\mathbb{D}$. We describe the decomposition of the tangent bundle of $\mathcal{T}(1)$ over the fiber $\pi^{-1}(0)$ and present isomorphisms between the tangent spaces. Lemma 1.5 which describes a special property of the quasiconformal mapping with harmonic Beltrami differential seems to be a new result. In Section 1.3 we present, in a succinct form, basic facts about the Teichmüller spaces and Teichmüller curves of Fuchsian groups, including the definition of the Weil-Petersson metric, and Patterson's lemma on the uniform distribution of lattice points on the hyperbolic plane. In Section 1.4 we collect necessary properties of the resolvent kernel $G = \frac{1}{2}(\Delta_0 + \frac{1}{2})^{-1}$ of the Laplace operator Δ_0 on the hyperbolic plane, and in Section 1.5 we present Ahlfors' classical variational formulas. In Section 2 we introduce new Hilbert manifold structure on $T(1)$. Namely, in Section 2.1 we define the Hilbert subspaces $H^{-1,1}(\mathbb{D}^*)$ and $A_2(\mathbb{D})$ of the tangent spaces to $T(1)$ and to $A_\infty(\mathbb{D})$. In Theorem 2.3 we prove that the differential of the Bers embedding $\beta : T(1) \to A_\infty(\mathbb{D})$ is a bounded bijection between these Hilbert spaces. In Section 2.2 we prepare all L^2-estimates used in Section 2.3. The main result there is Theorem 2.10 — the existence of a Hilbert manifold atlas for $T(1)$. In Theorem 2.13 we prove that the Bers embedding is also a biholomorphic mapping of Hilbert manifolds. In Section 3 we prove that $T_0(1)$ and the corresponding Teichmüller curve $\mathcal{T}_0(1)$ — the inverse image of $T_0(1)$ under the projection π, are topological groups in the Hilbert manifold topology. Moreover, we show that $T_0(1)$ is the closure of $\mathrm{M\ddot{o}b}(S^1)\backslash\mathrm{Diff}_+(S^1)$ in $T(1)$ with respect to this topology, and prove that $T_0(1)$ is the inverse image of $\beta(T(1)) \cap A_2(\mathbb{D})$ under the Bers embedding. In Section 4.1, following [**Teo04**], we recall the definition of the Velling-Kirillov metric on the universal Teichmüller curve $\mathcal{T}(1)$ considered as a Banach manifold, and in

---

[2]It is explained in [**Nag92**] and [**NS95**] it what sense the mapping $\hat{\mathscr{P}}$ generalizes the classical period mapping of compact Riemann surfaces.

Section 4.2 we define the Weil-Petersson metric on the Hilbert manifold $T(1)$. In Section 5.1 we prove that Velling-Kirillov metric is real-analytic on $\mathcal{T}(1)$ by explicitly constructing its real-analytic Kähler potential — Theorem 5.3. We introduce Mumford-Miller-Morita characteristic forms by considering $\pi : \mathcal{T}(1) \to T(1)$ as a fibration of Hilbert manifolds. The latter property is crucial for the operation "integration over the fibers" (which are non-compact) to be well-defined. In Theorem 5.10 we explicitly compute Mumford-Miller-Morita forms in terms of the resolvent G. This is an infinite-dimensional generalization of Wolpert's result in [**Wol86**]. In Section 6 we give a simple derivation of the second variation of the hyperbolic metric — Proposition 6.3. In Section 7 we prove that the Weil-Petersson metric on $T(1)$ is Kähler and explicitly compute its Riemann and Ricci curvature tensors, showing that $T(1)$ is a Kähler-Einstein manifold. The main results there are Theorem 7.7 and 7.11. They are based on a more technical Proposition 7.2 and Lemma 7.8. In Section 8 we derive Wolpert's curvature formulas [**Wol86**] for finite-dimensional Teichmüller spaces from the corresponding "universal" curvature formulas for $T(1)$, obtained in Section 7.

In Section 1 of Chapter 2 we characterize the univalent functions associated with the Hilbert manifold $T_0(1)$ in terms of the Hilbert spaces $A_2^1(\mathbb{D})$ and $A_2^1(\mathbb{D}^*)$ of holomorphic functions on $\mathbb{D}$ and $\mathbb{D}^*$ respectively, square integrable with respect to the Lebesgue measure. Using the embedding $A_2^1(\mathbb{D}) \hookrightarrow A_\infty^1(\mathbb{D})$ into the Banach space of holomorphic functions on $\mathbb{D}$, the Becker-Pommerenke theorem [**BP78**], and the characterization of the topological group S of symmetric homeomorphisms of S^1 given by Gardiner and Sullivan [**GS92**], we prove that $T_0(1)$ is a subgroup of S. The main result of this section is Theorem 1.12, which states that $[\mu] \in T_0(1)$ if and only if one of the following conditions holds: (i) $\mathcal{S}(f^\mu) \in A_2(\mathbb{D})$; (ii) $\mathcal{A}(f^\mu) \in A_2^1(\mathbb{D})$; (iii) $\mathcal{S}(g_\mu) \in A_2(\mathbb{D}^*)$; (iv) $\mathcal{A}(g_\mu) \in A_2^1(\mathbb{D}^*)$. Here $\mathcal{S}(f)$ is the Schwarzian derivative of the univalent function f, and

$$\mathcal{A}(f) = \frac{f''}{f'}.$$

This theorem allows us to introduce the "universal Liouville action" — the function $\mathsf{S}_1 : T_0(1) \to \mathbb{R}$, defined by

$$(0.1) \qquad \mathsf{S}_1([\mu]) = \iint_{\mathbb{D}} |\mathcal{A}(f^\mu)|^2 \, d^2z + \iint_{\mathbb{D}^*} |\mathcal{A}(g_\mu)|^2 \, d^2z - 4\pi \log |g'_\mu(\infty)|.$$

In Section 2 to every $[\mu] \in T(1)$ we assign the Grunsky operators B_1, B_2, B_3 and B_4, associated with the corresponding pair (f^μ, g_μ) of univalent functions. The Lebesgue measure of the quasi-circle $\mathbb{C} \setminus \{f(\mathbb{D}) \cup g(\mathbb{D}^*)\}$ is zero, so that the generalized Grunsky inequality [**Hum72, Pom92**] can be succinctly formulated as the unitarity of the operator $\mathbf{B} = \begin{pmatrix} B_1 & B_2 \\ B_3 & B_4 \end{pmatrix}$ on $\ell^2 \oplus \ell^2$. The main result of Section 2.1 is Theorem 2.6, which states (see Corollary 2.9) that $[\mu] \in T_0(1)$ if and only if the corresponding Grunsky operators $B_1(f^\mu), B_4(g_\mu) \in \mathscr{S}_2$ — the Hilbert space of Hilbert-Schmidt operators on ℓ^2. In Theorem 2.10 we prove that the mapping $\mathscr{P} : T_0(1) \to \mathscr{S}_2$, defined by $\mathscr{P}([\mu]) = B_1(f^\mu)$, is a holomorphic mapping of Hilbert manifolds. Extended to the universal Teichmüller space $T(1)$, this defines, as we prove in Appendix B, a holomorphic mapping $\hat{\mathscr{P}} : T(1) \to \mathscr{B}(\ell^2)$ of Banach manifolds. In Section 2.2 we show that for $[\mu] \in T_0(1)$ the eigenvalues of the corresponding trace class operators $B_1 B_1^*$ and $B_4 B_4^*$ are related to the eigenvalues of the classical Poincaré-Fredholm integral operator associated with the quasi-circle $\mathcal{C} = f^\mu(S^1) = g_\mu(S^1)$. Since for $[\mu] \in T_0(1)$ these quasi-circles contain all C^3 curves, this generalizes Schiffer's result [**Sch81**]. Extending [**Sch59**], we introduce the Fredholm determinant $\mathrm{Det}_F(\mathcal{C})$ of the quasi-circle $\mathcal{C}$ as the Fredholm determinant $\det(I - B_1 B_1^*) = \det(I - B_4 B_4^*)$, and define the function $\mathsf{S}_2 : T_0(1) \to \mathbb{R}$ by

$$(0.2) \qquad \mathsf{S}_2([\mu]) = \log \mathrm{Det}_F(f^\mu(S^1)), \quad [\mu] \in T(1).$$

In Section 2.3 we define the semi-infinite period matrices of 1-forms for natural bases of $A_2^1(\mathbb{D})$ and $A_2^1(\mathbb{D}^*)$, which generalize imaginary parts of the classical period matrices for compact Riemann surfaces, and show that they correspond to the operators $B_2 B_2^*$ and $B_3 B_3^*$.

In Section 3 we compute the "first variations" of the functions S_1 and S_2 — the $(1,0)$-forms $\partial \mathsf{S}_1$ and $\partial \mathsf{S}_2$, where ∂ is the $(1,0)$-component of the de Rham differential on the Hilbert manifold $T_0(1)$. Namely, we show in Theorems 3.5 and 3.1 (see Corollaries 3.9 and 3.2) that

$$(0.3) \qquad \partial \mathsf{S}_1 = 2\vartheta \quad \text{and} \quad \partial \mathsf{S}_2 = -\frac{1}{6\pi}\vartheta,$$

where the $(1,0)$-form ϑ on $T_0(1)$, under the natural isomorphism $T^*_{[\mu]} T_0(1) \simeq A_2(\mathbb{D}^*)$, is given by

$$(0.4) \qquad \vartheta_{[\mu]} = \mathcal{S}(g_\mu).$$

The proof of Theorem 3.1 is rather standard, whereas the proof of Theorem 3.5 relies heavily on the identity given in Lemma 3.6. The latter can be interpreted as an extension of the generalized Grunsky equality to pairs of univalent functions (f^μ, g_μ) for $[\mu] \in T_0(1)$, which we consider quite interesting. Since the functions S_1 and S_2 on $T_0(1)$ both vanish at $0 \in T_0(1)$, from (1.3) we immediately obtain that

$$\mathsf{S}_2 = -\frac{1}{12\pi}\mathsf{S}_1,$$

thus expressing the Fredholm determinant as the universal Liouville action. In Corollary 3.12 and Remark 3.13 we interpret this relation as a surgery type formula for the determinants of elliptic operators on domains on the Riemann sphere $\mathbb{P}^1$.

In Section 4 we show that the relation (1.3) implies that the function S_1 is a Kähler potential of the Weil-Petersson metric on $T_0(1)$. The proof goes along the same lines as in the case of finite-dimensional Teichmüller spaces [**TT03**]. This explains why the function S_1 is called the universal Liouville action. In Section 5 we study the period mapping $\hat{\mathscr{P}} : T(1) \to \mathscr{B}(\ell^2)$. We prove that it coincides with the Kirillov-Yuriev-Nag-Sullivan mapping of $T(1)$ into the infinite-dimensional analog of Siegel disk $\mathfrak{D}_\infty$. We also show that the period mapping $\mathscr{P} : T_0(1) \to \mathscr{S}_2$ gives an embedding of $T_0(1)$ into the Segal-Wilson universal Grassmannian.

In Appendix A we study the Hilbert manifold structure on the topological group $\mathcal{T}_0(1)$ — the pre-image of the Hilbert manifold $T_0(1)$ under the canonical projection $\pi : \mathcal{T}(1) \to T(1)$. We prove in Theorem A.3 that the Bers embedding $\beta : \mathcal{T}_0(1) \to A_2(\mathbb{D}) \oplus \mathbb{C}$ and the pre-Bers embedding $\hat{\beta} : \mathcal{T}_0(1) \to A_2^1(\mathbb{D})$ induce the same Hilbert manifold structure on $\mathcal{T}_0(1)$. This result is parallel to the one proved in the Appendix of [**Teo04**]. We also prove Corollaries A.4 and A.6, characterizing convergence in the Hilbert manifold topology of $\mathcal{T}_0(1)$, which were used in the proof of Lemma 4.6. Finally, in Appendix B we show that $\hat{\mathscr{P}} : T(1) \to \mathscr{B}(\ell^2)$ is a holomorphic mapping of Banach manifolds and prove that the image of the topological group S under the map $\hat{\mathscr{P}}$ is the subset $\mathscr{S}_\infty \cap \hat{\mathscr{P}}(T(1))$ of $\mathscr{B}(\ell^2)$. The properties of the tower of embedded manifolds $T_0(1) \hookrightarrow S \hookrightarrow T(1)$ are summarized in a commutative diagram at the end of Appendix B.

Acknowledgments. We appreciate useful discussions with C. Bishop and especially with M. Luybich. The second author would like to thank P.Y. Wu for helpful discussions about operator theory. The work of the first author was partially supported by the NSF grant DMS-0204628. The work of the second author was partially supported by the grants NSC 91-2115-M-009-017 and NSC 92-2115-M-009-017. The second author also thanks CTS for the fellowship to visit Stony Brook University in the Summer of 2003, where a part of this work was done.

CHAPTER 1

Curvature Properties and Chern Forms

1. The universal Teichmüller space

1.1. Teichmüller theory. Here we present, in a succinct form, necessary facts from Teichmüller theory (for more details, see monographs [**Ahl87, Leh87, Nag88**] and the exposition in [**Teo04**]).

1.1.1. *Main definitions.* Let $\mathbb{D} = \{z \in \mathbb{C} : |z| < 1\}$ be the open unit disk and let $\mathbb{D}^* = \{z \in \mathbb{C} : |z| > 1\}$ be its exterior. Denote by $L^\infty(\mathbb{D}^*)$ and $L^\infty(\mathbb{D})$ the complex Banach spaces of bounded Beltrami differentials on $\mathbb{D}^*$ and $\mathbb{D}$ respectively, and let $L^\infty(\mathbb{D}^*)_1$ be the open unit ball in $L^\infty(\mathbb{D}^*)$. Two classical models of the universal Teichmüller space $T(1)$ are the following.

Model A. Extend every $\mu \in L^\infty(\mathbb{D}^*)_1$ to $\mathbb{D}$ by the reflection

$$(1.1) \qquad \mu(z) = \overline{\mu\left(\frac{1}{\bar{z}}\right)} \frac{z^2}{\bar{z}^2} \, , \; z \in \mathbb{D},$$

and consider the unique quasiconformal (q.c.) mapping $w_\mu : \mathbb{C} \to \mathbb{C}$, which fixes $-1, -i$ and 1 (i.e., is normalized) and satisfies the Beltrami equation

$$(w_\mu)_{\bar{z}} = \mu(w_\mu)_z \, .$$

Here and in what follows subscripts z and $\bar{z}$ always stand for the partial derivatives $\frac{\partial}{\partial z}$ and $\frac{\partial}{\partial \bar{z}}$, unless it is explicitly stated otherwise. Due to the reflection symmetry (1.1) the q.c. mapping w_μ satisfies

$$(1.2) \qquad \frac{1}{\overline{w_\mu(z)}} = w_\mu\left(\frac{1}{\bar{z}}\right)$$

and fixes domains $\mathbb{D}, \mathbb{D}^*$, and the unit circle S^1. For $\mu, \nu \in L^\infty(\mathbb{D}^*)_1$, set $\mu \sim \nu$ if $w_\mu|_{S^1} = w_\nu|_{S^1}$. The universal Teichmüller space $T(1)$ is defined as a set of equivalence classes of normalized q.c. mappings w_μ,

$$T(1) = L^\infty(\mathbb{D}^*)_1 / \sim \, .$$

Model B. Extend every $\mu \in L^\infty(\mathbb{D}^*)_1$ to be zero outside $\mathbb{D}^*$, and consider the unique q.c. mapping w^μ which satisfies the Beltrami equation

$$w^\mu_{\bar{z}} = \mu w^\mu_z,$$

and is normalized by the conditions $f(0) = 0$, $f'(0) = 1$ and $f''(0) = 0$. Here $f = w^\mu|_\mathbb{D}$ is holomorphic on $\mathbb{D}$ and prime stands for the derivative. For $\mu, \nu \in L^\infty(\mathbb{D}^*)_1$, set $\mu \sim \nu$ if $w^\mu|_\mathbb{D} = w^\nu|_\mathbb{D}$. The universal Teichmüller space $T(1)$ is defined as a set of equivalence classes of normalized q.c. mappings w^μ,

$$T(1) = L^\infty(\mathbb{D}^*)_1 / \sim \, .$$

Since $w_\mu|_{S^1} = w_\nu|_{S^1}$ if and only if $w^\mu|_\mathbb{D} = w^\nu|_\mathbb{D}$, these two definitions of the universal Teichmüller space are equivalent. The set $T(1)$ is a topological space with the quotient topology induced from $L^\infty(\mathbb{D}^*)_1$. Denote by $\mathcal{L}^\infty(\mathbb{D}^*)$ the subspace of $L^\infty(\mathbb{D}^*)$ consisting of real-analytic Beltrami differentials. Every point in $T(1)$ can be represented by $\mu \in \mathcal{L}^\infty(\mathbb{D}^*)$ [**Leh87**, Sect. III.1.1].

The space $T(1)$ has a unique structure of a complex Banach manifold, such that the projection map
$$\Phi : L^\infty(\mathbb{D}^*)_1 \to T(1)$$
is a holomorphic submersion. The differential of Φ at the origin
$$D_0\Phi : L^\infty(\mathbb{D}^*) \to T_0 T(1)$$
is a complex linear surjection of holomorphic tangent spaces. The kernel of $D_0\Phi$ is the subspace $\mathcal{N}(\mathbb{D}^*)$ of infinitesimally trivial Beltrami differentials. Explicitly,
$$\mathcal{N}(\mathbb{D}^*) = \left\{ \mu \in L^\infty(\mathbb{D}^*) : \iint_{\mathbb{D}^*} \mu \phi \, d^2 z = 0 \text{ for all } \phi \in A_1(\mathbb{D}^*) \right\},$$
where $d^2 z = dx \wedge dy$, $z = x + iy$, and
$$A_1(\mathbb{D}^*) = \left\{ \phi \text{ holomorphic on } \mathbb{D}^* : \iint_{\mathbb{D}^*} |\phi| d^2 z < \infty \right\}.$$

The Banach space of bounded harmonic Beltrami differentials on $\mathbb{D}^*$ is defined by
$$\Omega^{-1,1}(\mathbb{D}^*) = \left\{ \mu \in L^\infty(\mathbb{D}^*) : \mu(z) = (1 - |z|^2)^2 \overline{\phi(z)}, \ \phi \in A_\infty(\mathbb{D}^*) \right\},$$
where
$$A_\infty(\mathbb{D}^*) = \left\{ \phi \text{ holomorphic on } \mathbb{D}^* : \|\phi\|_\infty = \sup_{z \in \mathbb{D}^*} \left|(1 - |z|^2)^2 \phi(z)\right| < \infty \right\}.$$

The Banach space $\Omega^{-1,1}(\mathbb{D}^*)$ is not separable. The decomposition
$$(1.3) \qquad L^\infty(\mathbb{D}^*) = \mathcal{N}(\mathbb{D}^*) \oplus \Omega^{-1,1}(\mathbb{D}^*)$$
identifies the holomorphic tangent space $T_0 T(1) = L^\infty(\mathbb{D}^*)/\mathcal{N}(\mathbb{D}^*)$ at the origin of $T(1)$ with the Banach space $\Omega^{-1,1}(\mathbb{D}^*)$. The universal Teichmüller space $T(1)$ is a complex Banach manifold modeled on $\Omega^{-1,1}(\mathbb{D}^*)$.

REMARK 1.1. Traditionally, the universal Teichmüller space is defined using the complex Banach space $L^\infty(\mathbb{D})_1$. The reflection (1.1) establishes natural complex anti-linear isomorphism between $L^\infty(\mathbb{D}^*)_1$ and $L^\infty(\mathbb{D})_1$, and the universal Teichmüller space in the traditional definition is complex conjugate to the space $T(1)$ defined above.

1.1.2. *The group structure.* The unit ball $L^\infty(\mathbb{D}^*)_1$ carries a group structure induced by the composition of q.c. mappings. The group law
$$\lambda = \nu * \mu^{-1}$$
is defined through $w_\lambda = w_\nu \circ w_\mu^{-1}$, where μ^{-1} stands for the inverse element to μ, i.e., $\mu * \mu^{-1} = 0$. The group law is given explicitly by
$$\lambda = \left(\frac{\nu - \mu}{1 - \bar{\mu}\nu} \frac{(w_\mu)_z}{\overline{(w_\mu)_{\bar{z}}}} \right) \circ w_\mu^{-1}.$$
It follows from this formula that $\mathcal{L}^\infty(\mathbb{D}^*)_1$ is a subgroup of $L^\infty(\mathbb{D}^*)_1$.

For every $\lambda \in L^\infty(\mathbb{D}^*)_1$ set $[\lambda] = \Phi(\lambda) \in T(1)$. The group structure on $L^\infty(\mathbb{D}^*)_1$ projects to $T(1)$ by $[\lambda] * [\mu] = [\lambda * \mu]$. For every $\mu \in L^\infty(\mathbb{D}^*)_1$ the right translations

$$R_{[\mu]} : T(1) \to T(1), \quad [\lambda] \mapsto [\lambda * \mu],$$

are biholomorphic automorphisms of $T(1)$. The left translations, in general, are not even continuous mappings (see, e.g., [**Leh87**, Sect. III.3.4]). For every $\mu \in L^\infty(\mathbb{D}^*)_1$ the kernel of $D_\mu \Phi$ is the subspace $D_0 R_\mu(\mathcal{N}(\mathbb{D}^*))$ of $L^\infty(\mathbb{D}^*)$ and

$$T_{[\mu]} T(1) = D_0 R_{[\mu]} (T_0 T(1)) \simeq D_0 R_\mu \left(\Omega^{-1,1}(\mathbb{D}^*) \right).$$

1.1.3. *The Bers embedding.* Let $A_\infty(\mathbb{D})$ be the complex Banach space

$$A_\infty(\mathbb{D}) = \left\{ \phi \text{ holomorphic on } \mathbb{D} : \|\phi\|_\infty = \sup_{z \in \mathbb{D}} \left| (1 - |z|^2)^2 \phi(z) \right| < \infty \right\}.$$

The Bers embedding $\beta : T(1) \hookrightarrow A_\infty(\mathbb{D})$ is defined as follows. Denote by $\mathcal{S}(f)$ the Schwarzian derivative of a conformal map f,

$$\mathcal{S}(f) = \frac{f_{zzz}}{f_z} - \frac{3}{2} \left(\frac{f_{zz}}{f_z} \right)^2.$$

For every $\mu \in L^\infty(\mathbb{D}^*)_1$ the holomorphic function $\mathcal{S}(w^\mu|_\mathbb{D}) \in A_\infty(\mathbb{D})$ (by Kraus-Nehari inequality it lies in the ball of radius 6 in $A_\infty(\mathbb{D})$). Set

$$\beta([\mu]) = \mathcal{S}(w^\mu|_\mathbb{D}).$$

The Bers embedding is a holomorphic map of complex Banach manifolds, and its differential at the origin is

(1.4) $$D_0 \beta(\mu)(z) = -\frac{6}{\pi} \iint_{\mathbb{D}^*} \frac{\mu(\zeta)}{(\zeta - z)^4} d^2\zeta.$$

The complex-linear mapping $D_0 \beta$ induces the isomorphism $\Omega^{-1,1}(\mathbb{D}^*) \xrightarrow{\sim} A_\infty(\mathbb{D})$ of the holomorphic tangent spaces to $T(1)$ and $A_\infty(\mathbb{D})$ at the origin. The mapping $\Lambda : A_\infty(\mathbb{D}) \to \Omega^{-1,1}(\mathbb{D}^*)$, inverse to $D_0 \beta$, is given by

$$\mu(z) = \Lambda(\phi)(z) = -\tfrac{1}{2}(1 - |z|^2)^2 \phi\left(\tfrac{1}{\bar{z}}\right) \tfrac{1}{\bar{z}^4}.$$

According to the Ahlfors-Weill theorem, over the ball of radius 2 in $A_\infty(\mathbb{D})$ the map $\phi \mapsto [\Lambda(\phi)]$ is the right inverse to β, $\beta \circ \Lambda = \mathrm{id}$.

1.1.4. *The complex structure.* For every $\mu \in L^\infty(\mathbb{D}^*)_1$ let $U_\mu \subset T(1)$ be the image of the ball of radius 2 in $A_\infty(\mathbb{D})$ under the map $h_\mu^{-1} = \Phi \circ R_\mu \circ \Lambda$. The maps $h_{\mu\nu} = h_\mu \circ h_\nu^{-1} : h_\nu(U_\mu \cap U_\nu) \to h_\mu(U_\mu \cap U_\nu)$ are biholomorphic as functions on the Banach space $A_\infty(\mathbb{D})$. The structure of $T(1)$ as a complex Banach manifold modeled on the Banach space $A_\infty(\mathbb{D})$ is explicitly described by the complex-analytic atlas given by the open covering

$$T(1) = \bigcup_{\mu \in L^\infty(\mathbb{D}^*)_1} U_\mu$$

with coordinate maps h_μ and transition maps $h_{\mu\nu}$. The canonical projection $\Phi : L^\infty(\mathbb{D}^*)_1 \to T(1)$ is a holomorphic submersion and the Bers embedding $\beta : T(1) \to A_\infty(\mathbb{D})$ is a biholomorphic map with respect to this complex structure.

REMARK 1.2. Since every point $T(1)$ can be represented by a real-analytic Beltrami differential, it is sufficient to consider the atlas formed by the charts (U_μ, h_μ) with $\mu \in \mathcal{L}^\infty(\mathbb{D}^*)_1$.

Complex coordinates on $T(1)$ defined by the coordinate charts (U_μ, h_μ) are called Bers coordinates. For every $\nu \in \Omega^{-1,1}(\mathbb{D}^*)$ set $\phi = D_0\beta(\nu)$ and define a holomorphic vector field $\frac{\partial}{\partial \varepsilon_\nu}$ on U_0 by setting

$$Dh_0\left(\frac{\partial}{\partial \varepsilon_\nu}\right) = \phi$$

at all points in U_0 [1]. At every point $[\mu] \in U_0$, identified with the corresponding harmonic Beltrami differential μ, the vector field $\frac{\partial}{\partial \varepsilon_\nu}$ in terms of the Bers coordinates on U_μ corresponds to

$$\tilde{\phi} = D_\mu h_\mu \left(\frac{\partial}{\partial \varepsilon_\nu}\right) = \left(D_\mu h_\mu \left(D_\mu h_0\right)^{-1}\right)(\phi) = D_0\left(\beta \circ \Phi\right)\left(D_\mu R_\mu^{-1}(\Lambda(\phi))\right).$$

Using identification $\Omega^{-1,1}(\mathbb{D}^*) \simeq A_\infty(\mathbb{D})$, provided by the mapping $D_0\beta$, we get

(1.5) $$\left.\frac{\partial}{\partial \varepsilon_\nu}\right|_\mu = P\left(D_\mu R_\mu^{-1}(\nu)\right) = P(R(\nu, \mu)),$$

where

(1.6) $$R(\nu, \mu) = \left(\frac{\nu}{1 - |\mu|^2} \frac{(w_\mu)_z}{(\overline{w}_\mu)_{\bar{z}}}\right) \circ w_\mu^{-1},$$

and $P : L^\infty(\mathbb{D}^*) \to \Omega^{-1,1}(\mathbb{D}^*)$ is the projection onto the subspace of harmonic Beltrami differentials, defined by the decomposition (1.3). Explicitly,

(1.7) $$(P\mu)(z) = \frac{3(1-|z|^2)^2}{\pi} \iint_{\mathbb{D}^*} \frac{\mu(\zeta)}{(1-\zeta\bar{z})^4} d^2\zeta.$$

REMARK 1.3. Right translating $\nu \in T_0 T(1)$ defines a holomorphic tangent vector

$$D_0 R_{[\mu]}(\nu) = (1 - |\mu|^2) \nu \circ w_\mu \frac{(\overline{w}_\mu)_{\bar{z}}}{(w_\mu)_z} \in T_{[\mu]}T(1)$$

at every $[\mu] \in T(1)$. In Bers coordinates on U_μ this tangent vector is represented by $\nu \in \Omega^{-1,1}(\mathbb{D}^*)$. However, the family $\{D_0 R_{[\mu]}(\nu)\}_{[\mu] \in T(1)}$ of holomorphic tangent vectors does not form a smooth vector field on $T(1)$ since the left translations are not continuous on $T(1)$.

1.1.5. *The universal Teichmüller curve.* The universal Teichmüller curve $\mathcal{T}(1)$ is a natural complex fiber space over $T(1)$ with a holomorphic projection map $\pi : \mathcal{T}(1) \to T(1)$. The fiber over each point $[\mu]$ is a quasi-disk $w^\mu(\mathbb{D}^*) \subset \hat{\mathbb{C}}$ with complex structure induced from $\hat{\mathbb{C}}$ and

(1.8) $$\mathcal{T}(1) = \{([\mu], z) : [\mu] \in T(1),\ z \in w^\mu(\mathbb{D}^*)\}.$$

The fibration $\pi : \mathcal{T}(1) \to T(1)$ has a natural holomorphic section given by $T(1) \ni [\mu] \mapsto ([\mu], \infty) \in \mathcal{T}(1)$ — the "zero section", which defines the embedding $T(1) \hookrightarrow \mathcal{T}(1)$. The universal Teichmüller curve is a complex Banach manifold modeled on $A_\infty(\mathbb{D}) \oplus \mathbb{C}$ [2], and the mapping

$$T(1) \times \mathbb{D}^* \ni ([\mu], z) \mapsto ([\mu], w^\mu(z)) \in \mathcal{T}(1)$$

is a real-analytic isomorphism.

---

[1] We identify holomorphic tangent space to $A_\infty(\mathbb{D})$ at every point with $A_\infty(\mathbb{D})$.
[2] Here $\hat{\mathbb{C}} \setminus \{0\}$ is identified with $\mathbb{C}$ via the conformal map $z \mapsto 1/z$.

1.2. Homogeneous spaces of Homeo$_{qs}(S^1)$. Let Homeo$_{qs}(S^1)$ be the group of orientation preserving quasi-symmetric homeomorphisms of the unit circle S^1 (see, e.g., [**Leh87**] for the definition), and let Diff$_+(S^1)$, Möb(S^1), and S^1 be the subgroups of Homeo$_{qs}(S^1)$ consisting, respectively, of smooth orientation preserving diffeomorphisms of S^1, of Möbius transformations of S^1, and of rotations of S^1.

Denote by $\mathcal{U}$ the set of univalent functions on $\mathbb{D}$ and let

$$\mathcal{D} = \{f \in \mathcal{U} : f(0) = 0, f'(0) = 1, f''(0) = 0, f \text{ admits a q.c. extension to } \mathbb{C}\},$$

$$\widetilde{\mathcal{D}} = \{f \in \mathcal{U} : f(0) = 0, f'(0) = 1, f \text{ admits a q.c. extension to } \mathbb{C}\}.$$

According to the Beurling-Ahlfors extension theorem, the maps

$$T(1) \ni [\mu] \mapsto w^\mu|_\mathbb{D} \in \mathcal{D}$$

and

$$T(1) \ni [\mu] \mapsto w_\mu|_{S^1} \in \text{Möb}(S^1)\backslash\text{Homeo}_{qs}(S^1)$$

define bijections

(1.9) $$\mathcal{D} \xleftarrow{\sim} T(1) \xrightarrow{\sim} \text{Möb}(S^1)\backslash\text{Homeo}_{qs}(S^1),$$

which endow the spaces $\mathcal{D}$ and Möb$(S^1)\backslash$Homeo$_{qs}(S^1)$ with the structure of complex Banach manifolds modeled on the Banach space $A_\infty(\mathbb{D})$. We will always identify the coset space Möb$(S^1)\backslash$Homeo$_{qs}(S^1)$ with the subgroup of Homeo$_{qs}(S^1)$ fixing $-1, -i$ and 1, so that the bijection $T(1) \xrightarrow{\sim} \text{Möb}(S^1)\backslash\text{Homeo}_{qs}(S^1)$ is a group isomorphism.

REMARK 1.4. It is a non-trivial problem to describe the complex Banach manifold structure of the spaces $\mathcal{D}$ and Möb$(S^1)\backslash$Homeo$_{qs}(S^1)$ intrinsically, without using the bijection (1.9).

1.2.1. *Conformal welding.* According to Beurling-Ahlfors extension theorem, for every $g \in \text{Möb}(S^1)\backslash\text{Homeo}_{qs}(S^1)$ there exists a unique $\alpha \in \text{Möb}(S^1)$ which fixes 1, and univalent functions f and g on $\mathbb{D}$ and $\mathbb{D}^*$, satisfying the following properties.

CW1. f and g admit q.c. extensions to $\mathbb{C}$.
CW2. $\alpha \circ \text{g} = (g^{-1} \circ f)|_{S^1}$.
CW3. $f(0) = 0, f'(0) = 1, f''(0) = 0$.
CW4. $g(\infty) = \infty$.

The factorization **CW2** is known as conformal welding. For $\text{g} = w_\mu|_{S^1}$, $[\mu] \in T(1)$, $f = w^\mu|_\mathbb{D}$ and $g = (w^\mu \circ w_\mu^{-1} \circ \alpha^{-1})|_{\mathbb{D}^*}$, so that $g(\mathbb{D}^*) = w^\mu(\mathbb{D}^*)$. Here w^μ is normalized so that f satisfies **CW3** and $\alpha \in \text{Möb}(S^1)$ is uniquely determined by the conditions $\alpha(1) = 1$ and **CW4**. For $[\mu] \in T(1)$ we will always denote $\text{g}_\mu = (\alpha \circ w_\mu)|_{S^1}$, $f^\mu = f$ and $g_\mu = g$, so that

$$\text{g}_\mu = (g_\mu^{-1} \circ f^\mu)|_{S^1}.$$

Slightly abusing notations, we will denote by g_μ a q.c. extension of $\text{g}_\mu = (\alpha \circ w_\mu)|_{S^1} \in \text{Möb}(S^1)\backslash\text{Homeo}_{qs}(S^1)$ given by $\alpha \circ w_\mu$. Since $\alpha \in \text{Möb}(S^1)$ fixes 1, the q.c. mapping g_μ satisfies the reflection property (1.2) and the factorization

(1.10) $$\text{g}_\mu = g_\mu^{-1} \circ f^\mu,$$

where $f^\mu = w^\mu$ and $g_\mu = w^\mu \circ w_\mu^{-1} \circ \alpha^{-1}$. We will distinguish between $\mathrm{g}_\mu \in \mathrm{Möb}(S^1)\backslash\mathrm{Homeo}_{qs}(S^1)$ and its q.c. extension by explicitly specifying either the property **CW2** or the factorization (1.10).

The following result will be used in Section 3.

LEMMA 1.5. *Let $\mu \in \Omega^{-1,1}(\mathbb{D}^*)_1 = \Omega^{-1,1}(\mathbb{D}^*) \cap L^\infty(\mathbb{D}^*)_1$ and $\mathrm{g}_\mu = \alpha \circ w_\mu$ the q.c. mapping introduced above. Then the mapping g_μ fixes 0 and ∞.*

PROOF. By the reflection property (1.2) and the factorization (1.10), it is sufficient to prove that $f^\mu = w^\mu$ fixes ∞. Denote

$$\gamma = \imath \circ \gamma_\mu \circ \imath, \quad \mathrm{g} = \imath \circ \mathrm{g}_\mu \circ \imath, \quad f = \imath \circ f^\mu \circ \imath, \quad \imath^*(\mu) = \mu \circ \imath \frac{\overline{\imath_z}}{\imath_z},$$

where $\imath(z) = z^{-1}$. The factorization (1.10) for g_μ gives $\mathrm{g} = g^{-1} \circ f$, and the property **CW3** for f^μ yields the following Laurent expansion of f at ∞,

$$(1.11) \qquad f(z) = z + \frac{a_1}{z} + \frac{a_2}{z^2} + \cdots.$$

We will prove that $f(0) = 0$ for $\mu \in \Omega^{-1,1}(\mathbb{D}^*)_1$ by exploiting the argument in Royden-Earle's proof of the Ahlfors-Weill theorem, as presented in [**Nag88**, Sect. 3.8.5].

Namely, f satisfies the Beltrami equation with the Beltrami differential $\nu = \imath^*(\mu)|_\mathbb{D}$, which is supported on $\mathbb{D}$. The fundamental theorem from the theory of q.c. mappings (see, e.g. [**Ahl87**]) asserts that f admits the series representation

$$(1.12) \qquad f(z) = z + P(\nu)(z) + P(\nu H(\nu))(z) + P(\nu H(\nu H(\nu)))(z) + \cdots,$$

which is uniformly and absolutely convergent on $\mathbb{C}$. Here for $h \in C^2(\mathbb{D})$ we denote

$$P(h)(z) = -\frac{1}{\pi} \iint_\mathbb{D} \frac{h(\zeta)}{\zeta - z} d^2\zeta,$$

$$H(h)(z) = -\frac{1}{\pi} \iint_\mathbb{D} \frac{h(\zeta)}{(\zeta - z)^2} d^2\zeta,$$

where the latter integral — the Hilbert transform, is understood in the principal value sense. Since ν has compact support, it immediately follows from the definition of the operators P and H that the series (1.12) has the Laurent expansion (1.11) at ∞. We will prove that for $\nu \in \Omega^{-1,1}(\mathbb{D})$ each term of this series vanishes at $z = 0$. Representing $\nu(z) = -\frac{1}{2}(1 - |z|^2)^2 \sum_{n=0}^\infty a_n \bar{z}^n$ and using polar coordinates, we get for any $(n-1)$ – iterate of the operator νH, $n > 1$,

$$P(\nu H(\nu H(\nu \ldots H(\nu))))(0)$$

$$= \left(\frac{1}{2\pi}\right)^n \iint_\mathbb{D} \cdots \iint_\mathbb{D} \frac{\sum_{m_1,\ldots m_n} a_{m_1} \ldots a_{m_n} r_1^{m_1+1} \ldots r_n^{m_n+1} e^{-im_1\theta_1} \ldots e^{-im_n\theta_n}}{r_1 e^{i\theta_1}(r_2 e^{i\theta_2} - r_1 e^{i\theta_1})^2 \ldots (r_n e^{i\theta_n} - r_{n-1} e^{i\theta_{n-1}})^2}$$

$$(1 - r_1^2)^2 \ldots (1 - r_n^2)^2 dr_1 d\theta_1 \ldots dr_n d\theta_n$$

$$= \sum_{m_1,\ldots,m_n=0}^\infty a_{m_1} \ldots a_{m_n} I_{m_1,\ldots,m_n},$$

where each integral in the definition of H is understood in the principal value sense. The interchange of the orders of summation and integration can be easily justified.

For fixed $r_1 \neq 0, r_1 \neq r_2, r_2 \neq r_3, \ldots, r_{n-1} \neq r_n$, let
$$I_{m_1,\ldots,m_n}(r_1,\ldots,r_n)$$
$$= \int_0^{2\pi} \cdots \int_0^{2\pi} \frac{e^{-im_1\theta_1}\cdots e^{-im_n\theta_n}d\theta_1 \cdots d\theta_n}{r_1 e^{i\theta_1}(r_2 e^{i\theta_2} - r_1 e^{i\theta_1})^2 \cdots (r_n e^{i\theta_n} - r_{n-1}e^{i\theta_{n-1}})^2}.$$

A change of variables $\theta_k \mapsto \theta_k + \theta$, $k=1,\ldots,n$ gives
$$I_{m_1,\ldots,m_n}(r_1,\ldots,r_n) = e^{-i(m_1+\ldots+m_n+(2n-1))\theta}I_{m_1,\ldots,m_n}(r_1,\ldots,r_n).$$

Since all $m_k \geq 0$ and $2n-1 > 0$ for $n \geq 1$, we have $e^{-i(m_1+\ldots+m_n+(2n-1))\theta} \neq 1$ and hence
$$I_{m_1,\ldots,m_n}(r_1,\ldots,r_n) = 0.$$

This proves that all $I_{m_1,\ldots,m_n}$ vanish. Therefore $f(0) = 0$. □

REMARK 1.6. Since $P(f)_z = H(f)$, it also follows from the proof that $f_z(0) = 1$.

Similar to (1.9), there are bijections
$$\widetilde{\mathcal{D}} \xleftarrow{\sim} \mathcal{T}(1) \xrightarrow{\sim} S^1 \backslash \mathrm{Homeo}_{qs}(S^1),$$
where we always identify the coset space $S^1\backslash\mathrm{Homeo}_{qs}(S^1)$ with the stabilizer of 1 in $\mathrm{Homeo}_{qs}(S^1)$ (see, e.g., [**Teo04**]). For every g $\in S^1\backslash\mathrm{Homeo}_{qs}(S^1)$ there exist unique univalent functions f and g on $\mathbb{D}$ and $\mathbb{D}^*$, satisfying the properties **CW1**, **CW4** and

CW2'. g $= (g^{-1}\circ f)|_{S^1}$;
CW3'. $f(0)=0$, $f'(0) = 1$.

Namely, the fibration $\pi : \mathcal{T}(1) \longrightarrow T(1)$ is the fiber space $S^1\backslash\mathrm{Homeo}_{qs}(S^1)$ over $\mathrm{M\ddot{o}b}(S^1)\backslash\mathrm{Homeo}_{qs}(S^1)$ with the fibers isomorphic to $S^1\backslash\mathrm{M\ddot{o}b}(S^1) \simeq \mathbb{D}^*$. The points in the fiber at $[\mu] \in T(1)$ correspond to the points $\sigma_w \circ \mathrm{g}_\mu \in S^1\backslash\mathrm{Homeo}_{qs}(S^1)$, $w \in \mathbb{D}^*$ with[3]
$$\sigma_w(z) = \frac{1-w}{1-\bar{w}}\frac{1-z\bar{w}}{z-w} \in S^1\backslash\mathrm{M\ddot{o}b}(S^1).$$

Using the properties **CW1** and **CW2** for g_μ, we get the factorization **CW2'** for γ is
$$\mathrm{g} = \sigma_w \circ \mathrm{g}_\mu = (g^{-1} \circ f)|_{S^1},$$
where
$$f = \lambda_w \circ f^\mu, \quad g = \lambda_w \circ g_\mu \circ \sigma_w^{-1},$$
and
$$\lambda_w(z) = \frac{z}{c_w z + 1}, \quad c_w = -\frac{1}{g_\mu(w)} = -\frac{1}{2}\frac{f''(0)}{f'(0)},$$
so that $(g_\mu \circ \sigma_w^{-1})(\infty) = g_\mu(w)$, and the functions f and g satisfy the properties **CW3'** and **CW4** respectively. The mapping
$$\mathcal{T}(1) \ni ([\mu], g_\mu(w)) \mapsto \mathrm{g} = \sigma_w \circ \mathrm{g}_\mu \in S^1\backslash\mathrm{Homeo}_{qs}(S^1)$$
establishes the isomorphism $\mathcal{T}(1) \xrightarrow{\sim} S^1\backslash\mathrm{Homeo}_{qs}(S^1)$.

As before, we will also denote by g a q.c. extension of g $\in S^1\backslash\mathrm{Homeo}_{qs}(S^1)$ which satisfies the reflection property (1.2) and admits the factorization g $= g^{-1} \circ f$.

---

[3]Here the subscript w does not stand for the derivative.

REMARK 1.7. It is known [**Kir87**] that $\mathrm{Diff}_+(S^1)$ is an infinite-dimensional Lie group and homogeneous spaces $\mathrm{Möb}(S^1)\backslash\mathrm{Diff}_+(S^1)$ and $S^1\backslash\mathrm{Diff}_+(S^1)$ are infinite-dimensional complex Fréchet manifolds. In this case conformal welding readily follows from the Riemann mapping theorem without using q.c. mappings [**Kir87**]. Note that our convention for the conformal welding is different from that in [**Kir87**]: we are using right cosets instead of left cosets.

The bijection $\mathcal{T}(1) \xrightarrow{\sim} S^1\backslash\mathrm{Homeo}_{qs}(S^1)$ endows the universal Teichmüller curve $\mathcal{T}(1)$ with the group structure. Explicitly,
$$([\lambda], z) = ([\nu], \zeta) * ([\mu], w)^{-1},$$
where
(1.13) $$\lambda = \left(\frac{\nu - \mu}{1 - \bar{\mu}\nu}\frac{\mathrm{g}_z}{\overline{\mathrm{g}_{\bar{z}}}}\right) \circ \mathrm{g}^{-1}$$
and
(1.14) $$z = \left(w^\lambda \circ \mathrm{g} \circ (w^\nu)^{-1}\right)(\zeta).$$

Here g is a q.c. extension of $\sigma_u \circ g_\mu$, $u = g_\mu^{-1}(w)$, and the point $([\lambda], z) \in \mathcal{T}(1)$ does not depend on the choice of the extension g.

The mapping
$$T(1) \ni [\mu] \mapsto \gamma_\mu \in S^1\backslash\mathrm{Homeo}_{qs}(S^1) \simeq \mathcal{T}(1)$$
is a complex–analytic embedding of $T(1)$ into $\mathcal{T}(1)$ which is not a group homomorphism. On the other hand, the natural embedding
$$\mathrm{Möb}(S^1)\backslash\mathrm{Homeo}_{qs}(S^1) \simeq T(1) \ni [\mu] \mapsto w_\mu \in S^1\backslash\mathrm{Homeo}_{qs}(S^1) \simeq \mathcal{T}(1)$$
is a group homomorphism, though not a complex–analytic mapping. Considering w_μ as an element of $S^1\backslash\mathrm{Homeo}_{qs}(S^1)$ via this embedding, it admits a conformal welding
(1.15) $$w_\mu = \mathrm{g}_\mu^{-1} \circ \mathrm{f}^\mu$$
that satisfies the properties **CW1, CW2′, CW3′, CW4**. In terms of the functions f^μ and g_μ satisfying properties **CW1–CW4**, we have
$$\mathrm{f}^\mu = \lambda_\mu \circ f^\mu \quad \text{and} \quad \mathrm{g}_\mu = \lambda_\mu \circ g_\mu \circ \alpha_\mu^{-1},$$
where $\alpha_\mu \in \mathrm{PSU}(1,1)$ is such that $w_\mu = \alpha_\mu \circ \gamma_\mu$, and $\lambda_\mu \in \mathrm{PSL}(2,\mathbb{C})$ is uniquely determined by the conditions $\mathrm{f}^\mu(0) = 0$, $(\mathrm{f}^\mu)'(0) = 1$ and $\mathrm{g}_\mu(\infty) = \infty$.

1.2.2. *The horizontal and vertical subspaces.* The right translations $R_{([\mu],z)} : \mathcal{T}(1) \to \mathcal{T}(1)$ are biholomorphic automorphisms of $\mathcal{T}(1)$ [**Ber73**]. The holomorphic tangent space to $\mathcal{T}(1)$ at $([\mu], z)$ is identified with the holomorphic tangent space at $(0, \infty)$ — the origin of $\mathcal{T}(1)$ by
$$T_{([\mu],z)}\mathcal{T}(1) = D_{(0,\infty)}R_{([\mu],z)}(T_{(0,\infty)}\mathcal{T}(1)) \simeq T_{(0,\infty)}\mathcal{T}(1).$$
The holomorphic tangent space at the origin naturally splits into the direct sum of horizontal and vertical subspaces,
$$T_{(0,\infty)}\mathcal{T}(1) = \Omega^{-1,1}(\mathbb{D}^*) \oplus \mathbb{C}.$$
The identification of holomorphic tangent spaces provides a natural splitting of the tangent space at every point in $\mathcal{T}(1)$ into the direct sum of horizontal and vertical subspaces. Lifts of horizontal and vertical tangent vectors at the origin of $\mathcal{T}(1)$ to every point in the fiber at the origin are explicitly described as follows.

TV1. Let $\mu \in \Omega^{-1,1}(\mathbb{D}^*) \subset T_{(0,\infty)}\mathcal{T}(1)$ be a horizontal tangent vector to $\mathcal{T}(1)$ at the origin. A curve $([t\mu], z(t))$, $z(0) = z$, which defines the horizontal lift of μ to the point $(0, z) \in \mathcal{T}(1)$ in the fiber $\pi^{-1}(0)$ at the origin, for small t is given by the equation

$$([\mu(t)], \infty) * (0, z) = ([t\mu], z(t)).$$

Using (1.13), (1.14) and Lemma 1.5, we get

$$\mu(t) = (\sigma_z^{-1})^*(t\mu) = t\mu \circ \sigma_z^{-1} \frac{\overline{(\sigma_z^{-1})'}}{(\sigma_z^{-1})'} \quad \text{and} \quad z(t) = w^{t\mu}(z).$$

Thus the horizontal lift of $\mu \in T_{(0,\infty)}\mathcal{T}(1)$ to every point in the fiber $(0, z) \in \pi^{-1}(0)$ is the vector field

$$\tau_\mu = \left.\frac{\partial}{\partial \varepsilon_\mu}\right|_0 + \dot{w}^\mu(z)\frac{\partial}{\partial z}, \quad \text{where} \quad \dot{w}^\mu(z) = \frac{dz}{dt}(0)$$

(cf. [**Wol86**]). At the point $(0, z) \in \pi^{-1}(0)$ the vector field τ_μ is identified with the horizontal tangent vector $(\sigma_z^{-1})^*\mu \in T_{(0,\infty)}\mathcal{T}(1)$.

TV2. Let $1 \in \mathbb{C} \subset T_{(0,\infty)}\mathcal{T}(1)$ be the vertical tangent vector to $\mathcal{T}(1)$ at the origin, given by the value of the vector field $\partial_z = \frac{\partial}{\partial z}$ at $z = \infty$. A curve defining the right translate of $\mathbf{1}$ to the point $(0, z) \in \mathcal{T}(1)$ in the fiber $\pi^{-1}(0)$ at the origin for small t is given by the equation

$$(0, t^{-1}) * (0, z) = (0, z(t)),$$

and it follows from (1.14) that

$$\frac{dz}{dt}(0) = \frac{(1-z)(1-|z|^2)}{(1-\bar{z})}.$$

Thus the right translate of $\mathbf{1} \in T_{(0,\infty)}\mathcal{T}(1)$ to the point $(0, z) \in \pi^{-1}(0)$ is the vector $\frac{(1-z)(1-|z|^2)}{(1-\bar{z})}\partial_z$ at $(0, z)$. As a result, the vector field ∂_z at every point $(0, z) \in \pi^{-1}(0)$ is identified with the vertical tangent vector

$$\frac{(1-\bar{z})}{(1-z)(1-|z|^2)}\mathbf{1} \in T_{(0,\infty)}\mathcal{T}(1).$$

1.2.3. *The isomorphisms of the tangent spaces.* The real tangent vector space $T_0^{\mathbb{R}} S^1\backslash \mathrm{Homeo}_{qs}(S^1)$ to $S^1\backslash \mathrm{Homeo}_{qs}(S^1)$ at the origin is identified with the subspace of Zygmund class continuous real-valued vector fields $\mathrm{u} = u(\theta)\frac{d}{d\theta}$ on S^1 (see, e.g., [**Teo04**] for the definition), satisfying

$$\int_0^{2\pi} u(\theta)d\theta = 0.$$

In particular, the Fourier series $u(\theta) = \sum_{n\in\mathbb{Z}} c_n e^{in\theta}$ is absolutely convergent. For $|z| = 1$ set

$$\tilde{u}(z) = i\sum_{n\in\mathbb{Z}\backslash\{0\}} c_n z^{n+1}.$$

The function $\tilde{u}$ on S^1 admits the decomposition

$$\tilde{u} = u_+ + u_-,$$

where u_+ and u_- are boundary values of functions holomorphic on $\mathbb{D}$ and $\mathbb{D}^*$ respectively and $u_+(0) = 0$. Explicitly,

$$u_+(z) = i \sum_{n=1}^{\infty} c_n z^{n+1},$$

$$u_-(z) = i \sum_{n=1}^{\infty} c_{-n} z^{1-n}.$$

It is a difficult problem to characterize the Zygmund class in terms of the Fourier series (cf. Remark 1.4). On the other side, in terms of the Fourier series the almost complex structure J on $T_0^{\mathbb{R}} S^1 \backslash \mathrm{Homeo}(S^1)$ is explicitly given by the classical conjugation operator

$$J\mathrm{u} = i \sum_{n \in \mathbb{Z} \backslash \{0\}} \mathrm{sgn}(n) c_n e^{in\theta} \frac{d}{d\theta} \quad \text{for} \quad \mathrm{u} = \sum_{n \in \mathbb{Z} \backslash \{0\}} c_n e^{in\theta} \frac{d}{d\theta}.$$

REMARK 1.8. Note that our definition of the operator J differs by a negative sign from the definition in [**Kir87**, **NV90**] for the homogeneous space $S^1 \backslash \mathrm{Diff}_+(S^1)$.

The holomorphic and anti-holomorphic tangent vectors at the origin are

$$\mathrm{v} = \frac{\mathrm{u} - iJ\mathrm{u}}{2} = \sum_{n=1}^{\infty} c_n e^{in\theta} \frac{d}{d\theta} \quad \text{and} \quad \bar{\mathrm{v}} = \frac{\mathrm{u} + iJ\mathrm{u}}{2} = \sum_{n=-\infty}^{-1} c_n e^{in\theta} \frac{d}{d\theta}.$$

For every smooth function $\mathcal{F}$ in a neighborhood of the origin in $\mathcal{T}(1)$ and $\mathrm{u} \in T_0^{\mathbb{R}} S^1 \backslash \mathrm{Homeo}_{qs}(S^1)$ set

$$\dot{\mathcal{F}}[\mathrm{u}] = \frac{d}{dt}\bigg|_{t=0} \mathcal{F}(\mathrm{g}^t),$$

where g_t is a curve in $S^1 \backslash \mathrm{Homeo}_{qs}(S^1)$ with the tangent vector u at the origin. Corresponding directional derivatives of $\mathcal{F}$ at the origin in $\mathcal{T}(1)$ in the holomorphic and anti-holomorphic directions v and $\bar{\mathrm{v}}$ are defined by

$$(1.16) \quad \partial \mathcal{F}(\mathrm{v}) = \frac{1}{2}\left(\dot{\mathcal{F}}[\mathrm{u}] - i \dot{\mathcal{F}}[J\mathrm{u}]\right), \quad \text{and} \quad \bar{\partial} \mathcal{F}(\bar{\mathrm{v}}) = \frac{1}{2}\left(\dot{\mathcal{F}}[\mathrm{u}] + i \dot{\mathcal{F}}[J\mathrm{u}]\right).$$

For $s \in \mathbb{R}$ let

$$\mathcal{H}^s(S^1) = \left\{ \mathrm{u} = \sum_{n=-\infty}^{\infty} a_n e^{in\theta} \frac{d}{d\theta} : \sum_{n=-\infty}^{\infty} |n|^{2s} |a_n|^2 < \infty \right\}$$

be the Sobolev space of complex-valued vector fields on S^1. The properties of the tangent spaces $T_0 S^1 \backslash \mathrm{Homeo}_{qs}(S^1)$, $T_0 \widetilde{\mathcal{D}}$ and $T_0 \mathrm{M\ddot{o}b}(S^1) \backslash \mathrm{Homeo}_{qs}(S^1)$, which will be used in Section 5, can be succinctly summarized as follows (see [**Teo04**] for details).

TS1. Under the $\mathbb{R}$-linear isomorphism $T_0^{\mathbb{R}} S^1 \backslash \mathrm{Homeo}_{qs}(S^1) \xrightarrow{\sim} T_0 \widetilde{\mathcal{D}}$

$$u(\theta) = \sum_{n \in \mathbb{Z} \backslash \{0\}} c_n e^{in\theta} \mapsto u_+(z) = i \sum_{n=1}^{\infty} c_n z^{n+1},$$

and $\dot{f}|_{\mathbb{D}} = u_+$, $\dot{g}_0|_{\mathbb{D}^*} = -u_-$, where

$$\dot{f} = \frac{d}{dt} f^t \bigg|_{t=0}, \quad \dot{g} = \frac{d}{dt} g_t \bigg|_{t=0},$$

$\gamma_t = g_t^{-1} \circ f^t$ is a smooth curve in $\mathcal{T}(1)$ tangent to u at the origin, and $\dot{g}_0(z) = \dot{g}(z) - \dot{g}'(\infty)z$.

TS2. Under the $\mathbb{R}$-linear isomorphism

$$T_0^{\mathbb{R}} \operatorname{M\ddot{o}b}(S^1)\backslash \operatorname{Homeo}_{qs}(S^1) \xrightarrow{\sim} T_0 T(1) \xrightarrow{D_0\beta} A_\infty(\mathbb{D})$$

$$u(\theta) = \sum_{n \in \mathbb{Z}\backslash\{-1,0,1\}} c_n e^{in\theta} \mapsto \frac{d^3 u_+}{dz^3}(z) = i\sum_{n=2}^\infty (n^3 - n)c_n z^{n-2}.$$

TS3. If

$$\phi(z) = \sum_{n=2}^\infty (n^3 - n)a_n z^{n-2} \in A_\infty(\mathbb{D})$$

then

$$\sum_{n=2}^\infty n^{2s}|a_n|^2 < \infty \quad \text{for all} \quad s < 1.$$

TS4. $T_0^{\mathbb{R}} S^1\backslash \operatorname{Homeo}_{qs}(S^1) \subset \mathcal{H}^s(S^1)$ for all $s < 1$.

1.3. Teichmüller spaces and Teichmüller curves of Fuchsian groups.
Let Γ be a Fuchsian group, i.e., a discrete subgroup of $\operatorname{PSU}(1,1)$. Let

$$L^\infty(\mathbb{D}^*, \Gamma) = \left\{\mu \in L^\infty(\mathbb{D}^*) : \mu \circ \gamma \frac{\overline{\gamma'}}{\gamma'} = \mu \quad \text{for all} \quad \gamma \in \Gamma\right\}$$

be the space of bounded Beltrami differentials for Γ and

$$L^\infty(\mathbb{D}^*, \Gamma)_1 = L^\infty(\mathbb{D}^*)_1 \cap L^\infty(\mathbb{D}^*, \Gamma)$$

be the open unit ball in $L^\infty(\mathbb{D}^*, \Gamma)$. The Teichmüller space of the Fuchsian group Γ is defined by

$$T(\Gamma) = L^\infty(\mathbb{D}^*, \Gamma)_1 / \sim,$$

where the equivalence relation is the same as the one used to define the universal Teichmüller space $T(1)$ in Section 1.1.1. The Teichmüller space $T(\Gamma)$ has a natural structure of a complex Banach manifold such that the tangent space at the origin of $T(\Gamma)$ is identified with the Banach space $\Omega^{-1,1}(\mathbb{D}^*, \Gamma)$ of bounded harmonic Beltrami differentials for Γ,

$$\Omega^{-1,1}(\mathbb{D}^*, \Gamma) = \Omega^{-1,1}(\mathbb{D}^*) \cap L^\infty(\mathbb{D}^*, \Gamma).$$

For every Fuchsian group Γ the canonical embedding $T(\Gamma) \hookrightarrow T(1)$ is holomorphic, so that the universal Teichmüller space $T(1)$ contains all the Teichmüller spaces $T(\Gamma)$ as complex submanifolds. The universal Teichmüller space $T(1)$ is the Teichmüller space for the trivial Fuchsian group $\Gamma = \{1\}$.

The inverse image of $T(\Gamma)$ under the projection map $\mathcal{T}(1) \to T(\Gamma)$ is called the Bers fiber space $\mathcal{BF}(\Gamma)$. The quasi-Fuchsian group $\Gamma^\mu = w^\mu \circ \Gamma \circ (w^\mu)^{-1}$ acts on the fiber $w^\mu(\mathbb{D}^*)$ at the point $[\mu] \in T(\Gamma)$. The Teichmüller curve $\mathcal{T}(\Gamma)$ of the Fuchsian group Γ is a fiber space over $T(\Gamma)$ with the fiber $\Gamma^\mu\backslash w^\mu(\mathbb{D}^*)$ at the point $[\mu] \in T(\Gamma)$.

The domain $\mathbb{D}^*$ is a model of the hyperbolic plane $\mathbb{H}^2$. The hyperbolic (Poincaré) metric on $\mathbb{D}^*$ — a Hermitian metric of constant Gaussian curvature -1, is

(1.17) $$ds^2 = \rho(z)|dz|^2 = \frac{4|dz|^2}{(1-|z|^2)^2},$$

and the hyperbolic area 2-form is $\rho(z)\,d^2z$. The Fuchsian group Γ is of finite type (cofinite) if the corresponding Riemann surface — the orbifold $\Gamma\backslash\mathbb{D}^*$, has a finite hyperbolic area. In this case, the Teichmüller space $T(\Gamma)$ is a finite-dimensional complex manifold with a natural Hermitian metric, called the Weil-Petersson metric. It is defined as Petersson's inner product on tangent spaces $T_{[\mu]}T(\Gamma) \simeq \Omega^{-1,1}(\mathbb{D}^*, \Gamma_\mu)$, where $[\mu] \in T(\Gamma)$ and $\Gamma_\mu = w_\mu \circ \Gamma \circ w_\mu^{-1}$. For $\mu, \nu \in T_0 T(\Gamma)$,

$$\langle \mu, \nu \rangle_{WP} = \iint_{\Gamma\backslash\mathbb{D}^*} \mu\bar\nu \rho(z) d^2 z.$$

The Weil-Petersson metric on $T(\Gamma)$ is a Kähler metric.

The following result, due to Patterson [**Pat75**], will be used in Section 8. Here we present it in a convenient form as in [**Teo04**].

LEMMA 1.9. *Let Γ be a cofinite Fuchsian group and $h \in L^\infty(\mathbb{D}^*, \rho(z)d^2z)$ be Γ-automorphic, i.e., $h \circ \gamma = h$ for all $\gamma \in \Gamma$. Then*

$$\iint_{\Gamma\backslash\mathbb{D}^*} h(z)\rho(z)d^2z = \lim_{r \to 1^+} \frac{A(\Gamma\backslash\mathbb{D}^*)}{A(\mathbb{D}_r^*)} \iint_{\mathbb{D}_r^*} h(z)\rho(z)d^2z,$$

where $\mathbb{D}_r^* = \{z \in \mathbb{D}^* : |z| \geq r\}$, $A(\Gamma\backslash\mathbb{D}^*)$ is the hyperbolic area of the Riemann surface $\Gamma\backslash\mathbb{D}^*$, and $A(\mathbb{D}_r^*)$ is the hyperbolic area of $\mathbb{D}_r^*$.

1.4. Resolvent kernel. Let

(1.18) $$\Delta_0 = -\rho(z)^{-1}\partial_z \partial_{\bar z}$$

be the Laplace-Beltrami operator of the hyperbolic metric on $\mathbb{D}$, acting on functions. It is well-known (see, e.g., [**Hej76, Lan85**]) that the differential expression (1.18) defines a unique positive, self-adjoint operator on the Hilbert space $L^2(\mathbb{D}, \rho(z)d^2z)$, which we still denote by Δ_0. Let

$$G = \tfrac{1}{2}\left(\Delta_0 + \tfrac{1}{2}\right)^{-1}$$

be (a one-half of) the resolvent of Δ_0 at the regular point $\lambda = -\tfrac{1}{2}$.

REMARK 1.10. Note that the Laplace-Beltrami operator in [**Hej76, Lan85**] is $4\Delta_0$, so that the regular point $\lambda = -\tfrac{1}{2}$ for the operator Δ_0 corresponds to $\lambda = -2$ for the Laplace-Beltrami operator in [**Hej76, Lan85**].

The resolvent G is a bounded integral operator on $L^2(\mathbb{D}, \rho(z)d^2z)$ with kernel

(1.19) $$G(z,w) = \frac{2u+1}{2\pi}\log\frac{u+1}{u} - \frac{1}{\pi},$$

where $u(z,w)$ is a point-pair invariant on $\mathbb{D}$,

$$u(z,w) = \frac{|z-w|^2}{(1-|z|^2)(1-|w|^2)}.$$

The resolvent kernel $G(z,w)$ has the following properties (see, e.g., [**Hej76**] and [**Lan85**, Sect. XIV.3]).

RK1. G is symmetric, $G(z,w) = G(w,z)$, and is a point-pair invariant,

$$G(\gamma z, \gamma w) = G(z,w) \text{ for all } \gamma \in \mathrm{PSU}(1,1).$$

RK2. $G(z,w)$ is positive for all $z, w \in \mathbb{D}$.

RK3. If $g \in BC^\infty(\mathbb{D})$ — the space of smooth bounded functions on $\mathbb{D}$, then the integral
$$f(z) = \iint_{\mathbb{D}} G(z,w)g(w)\rho(w)d^2w$$
is absolutely convergent for all $z \in \mathbb{D}$ and $f = G(g) \in BC^\infty(\mathbb{D})$ satisfies the differential equation
$$2\left(\Delta_0 + \tfrac{1}{2}\right)(f) = g.$$
Conversely, if $f \in BC^\infty(\mathbb{D})$ and $g = 2\left(\Delta_0 + \tfrac{1}{2}\right)(f) \in BC^\infty(\mathbb{D})$, then $f = G(g)$.

RK4. For all $z \in \mathbb{D}$,
$$\iint_{\mathbb{D}} G(z,w)\rho(w)d^2w = 1.$$

The last property immediately follows from **RK3** since
$$2\left(\Delta_0 + \tfrac{1}{2}\right)(1) = 1,$$
where 1 is the constant function equal to 1 on $\mathbb{D}$.

The resolvent kernel G of the Laplace-Beltrami operator on $\mathbb{D}^*$ is given by the same formula (1.19) and satisfies the properties **RK1** – **RK4**.

When Γ is a cofinite Fuchsian group, we denote by G_Γ the one-half of the resolvent of the Laplace-Beltrami operator on the Riemann surface $\Gamma\backslash\mathbb{D}$ at $\lambda = -\tfrac{1}{2}$. It is a bounded integral operator on $L^2(\Gamma\backslash\mathbb{D}, \rho(z)d^2z)$ with kernel
$$(1.20) \qquad G_\Gamma(z,w) = \sum_{\gamma \in \Gamma} G(z, \gamma w), \; z, w \in \mathbb{D},$$
and it enjoys all the properties **RK1**-**RK4**. The corresponding resolvent kernel on $\Gamma\backslash\mathbb{D}^*$ is given by the same formula with $z, w \in \mathbb{D}^*$.

REMARK 1.11. The operator G_Γ plays a prominent role in the Weil-Petersson geometry of the finite-dimensional Teichmüller space $T(\Gamma)$ [**Wol86**].

1.5. Variational formulas. Here we collect necessary variational formulas. To simplify the computations in the following sections, we will use different realizations of the hyperbolic plane $\mathbb{H}^2$, given either by the unit disk $\mathbb{D}$ or its exterior $\mathbb{D}^*$, or by the upper half-plane $\mathbb{U}$.

Let l and m be integers and Γ a Fuchsian group (we will be primarily interested in the cases when $\Gamma = \{1\}$, i.e., is a trivial group, and when Γ is a cofinite Fuchsian group). Using the model $\mathbb{H}^2 \simeq \mathbb{D}$, tensor of type (l, m) for Γ is a C^∞-function ω on $\mathbb{D}$ satisfying
$$\omega(\gamma z)\gamma'(z)^l \overline{\gamma'(z)}^m = \gamma(z) \text{ for all } \gamma \in \Gamma.$$
Let ω^ε be a smooth family of tensors of type (l, m) for $\Gamma_{\varepsilon\mu} = w_{\varepsilon\mu} \circ \Gamma \circ w_{\varepsilon\mu}^{-1}$, where $\mu \in \Omega^{-1,1}(\mathbb{D}, \Gamma)$ and $\varepsilon \in \mathbb{C}$ is sufficiently small. Set
$$(w_{\varepsilon\mu})^*(\omega^\varepsilon) = \omega^\varepsilon \circ w_{\varepsilon\mu}((w_{\varepsilon\mu})_z)^l((\overline{w}_{\varepsilon\mu})_{\bar{z}})^m,$$
which is a tensor of type (l, m) for Γ — a pull-back of the tensor ω^ε by $w_{\varepsilon\mu}$. Lie derivatives of the family ω^ε along vector fields $\partial/\partial\varepsilon_\mu$ and $\partial/\partial\bar\varepsilon_\mu$ are defined in the standard way,
$$L_\mu \omega = \left.\frac{\partial}{\partial\varepsilon}\right|_{\varepsilon=0}(w_{\varepsilon\mu})^*(\omega^\varepsilon) \quad \text{and} \quad L_{\bar\mu}\omega = \left.\frac{\partial}{\partial\bar\varepsilon}\right|_{\varepsilon=0}(w_{\varepsilon\mu})^*(\omega^\varepsilon).$$

When ω is a function on $T(\Gamma)$ — a tensor of type $(0,0)$, the Lie derivatives reduce to directional derivatives

$$L_\mu \omega = \partial \omega(\mu) \quad \text{and} \quad L_{\bar\mu} \omega = \bar\partial \omega(\bar\mu)$$

— the evaluation of the 1-forms $\partial \omega$ and $\bar\partial \omega$ on the holomorphic and antiholomorphic tangent vectors μ and $\bar\mu$ to $T(\Gamma)$ at the origin. Corresponding real vector fields $\frac{\partial}{\partial t_\mu}$ are defined by

$$\frac{\partial}{\partial t_\mu} = \frac{\partial}{\partial \varepsilon_\mu} + \frac{\partial}{\partial \bar\varepsilon_\mu},$$

so that

$$\frac{\partial}{\partial \varepsilon_\mu} = \frac{1}{2}\left(\frac{\partial}{\partial t_\mu} - i\frac{\partial}{\partial t_{i\mu}}\right) \quad \text{and} \quad \frac{\partial}{\partial \bar\varepsilon_\mu} = \frac{1}{2}\left(\frac{\partial}{\partial t_\mu} + i\frac{\partial}{\partial t_{i\mu}}\right).$$

For the model $\mathbb{H}^2 \simeq \mathbb{U}$ we have

(1.21) $$\frac{\partial}{\partial \varepsilon} w_{\varepsilon\mu}(z) = -\frac{1}{\pi} \iint_{\mathbb{U}} R\left(w_{\varepsilon\mu}(z), w_{\varepsilon\mu}(u)\right) \mu(u)(w_{\varepsilon\mu})_u^2(u) d^2u,$$

$$\frac{\partial}{\partial \bar\varepsilon} w_{\varepsilon\mu}(z) = -\frac{1}{\pi} \iint_{\mathbb{U}} R\left(w_{\varepsilon\mu}(z), \overline{w_{\varepsilon\mu}(u)}\right) \overline{\mu(u)(w_{\varepsilon\mu})_u^2(u)} d^2u,$$

where the q.c. mapping $w_{\varepsilon\mu}$ is normalized by fixing $0, 1, \infty$ and the kernel R is

$$R(z, u) = \frac{z(z-1)}{(u-z)u(u-1)} = \frac{1}{u-z} + \frac{z-1}{u} - \frac{z}{u-1}.$$

Setting

$$F[\mu] = \left.\frac{\partial}{\partial \varepsilon}\right|_{\varepsilon=0} w_{\varepsilon\mu} \quad \text{and} \quad \Phi[\mu] = \left.\frac{\partial}{\partial \bar\varepsilon}\right|_{\varepsilon=0} w_{\varepsilon\mu},$$

we get from (1.21)

(1.22) $$F[\mu](z) = -\frac{1}{\pi} \iint_{\mathbb{U}} R(z, u)\mu(u) \, d^2u,$$

$$\Phi[\mu](z) = -\frac{1}{\pi} \iint_{\mathbb{U}} R(z, \bar u)\overline{\mu(u)} \, d^2u.$$

The function $\Phi[\mu](z)$ is holomorphic on $\mathbb{U}$ and satisfies

$$\Phi[\mu]_{zzz}(z) = -\frac{6}{\pi} \iint_{\mathbb{U}} \frac{\overline{\mu(u)}}{(\bar u - z)^4} \, d^2u.$$

As it follows from (1.7), the projection $P: L^\infty(\mathbb{U}) \to \Omega^{-1,1}(\mathbb{U})$ is given by

(1.23) $$(P\mu)(z) = -\frac{3(z-\bar z)^2}{\pi} \iint_{\mathbb{U}} \frac{\mu(u)}{(u-\bar z)^4} d^2u.$$

Equivalently, for $\mu(z) = \frac{(z-\bar z)^2}{2}\overline{\phi(z)}$ with $\phi \in A_\infty(\mathbb{U})$, $\Phi[\mu]_{zzz} = \phi$ on $\mathbb{U}$. The function $F[\mu]$ satisfies $F[\mu]_{\bar z} = \mu$ on $\mathbb{U}$, and is holomorphic on the lower half-plane $\overline{\mathbb{U}}$.

LEMMA 1.12. *For $\mu \in \Omega^{-1,1}(\mathbb{U})$ and $z \in \mathbb{U}$,*

$$\lim_{\varepsilon \to 0} \iint_{\mathbb{U}(z,\varepsilon)} \frac{\mu(u)}{(u-z)^4} d^2u = \lim_{\varepsilon \to 0} \iint_{\mathbb{U}(z,\varepsilon)} \frac{\mu(u)}{(u-z)^5} d^2u = 0,$$

where $\mathbb{U}(z,\varepsilon) = \mathbb{U} \setminus \{u \in \mathbb{U} : |u - z| < \varepsilon\}$.

PROOF. The proof of the first formula essentially follows the classical Ahlfors' proof in [**Ahl87**, Lemma 2 in Sect. VI D] by using $\mu(u) = \frac{(u-\bar{u})^2}{2} \overline{\phi(u)}$ with $\phi \in A_\infty(\mathbb{U})$, the identity

$$\frac{(u-\bar{u})^2}{(u-z)^4} = \frac{\partial}{\partial u}\left(-\frac{1}{u-z} + \frac{\bar{u}-z}{(u-z)^2} - \frac{1}{3}\frac{(\bar{u}-z)^2}{(u-z)^3}\right),$$

and Stokes' theorem. The second formula is proved similarly. □

Another classical result of Ahlfors [**Ahl61**] is the following.

LEMMA 1.13. *For $\mu \in \Omega^{-1,1}(\mathbb{U})$ and $z \in \mathbb{U}$,*

$$F[\mu](z) = \frac{(z-\bar{z})^2}{2}\overline{\Phi''(z)} + (z-\bar{z})\overline{\Phi'(z)} + \overline{\Phi(z)},$$

where $\Phi(z) = \Phi[\mu](z)$.

REMARK 1.14. It follows from Lemma 1.13 that $F[\mu]_{zzz} = 0$ for $\mu \in \Omega^{-1,1}(\mathbb{U})$, in agreement with Lemma 1.12.

COROLLARY 1.15. *For $\mu \in \Omega^{-1,1}(\mathbb{U})$ and $z \in \mathbb{U}$,*

$$\iint_{\mathbb{U}} \frac{\mu(u)}{(u-z)(u-\bar{z})^3} d^2u = 0.$$

PROOF. Using (1.22), we have

$$F[\mu](z) - \frac{(z-\bar{z})^2}{2}\overline{\Phi''(z)} - (z-\bar{z})\overline{\Phi'(z)} - \overline{\Phi(z)}$$
$$= -\frac{1}{\pi}\iint_{\mathbb{U}} \mu(u)\frac{(z-\bar{z})^3}{(u-z)(u-\bar{z})^3} d^2w.$$

□

For $\mu \in L^\infty(\mathbb{U})_1$ set

$$K_\mu(u,v) = \frac{(w_\mu)_u(u)(w_\mu)_v(v)}{(w_\mu(u) - w_\mu(v))^2} \quad \text{and} \quad K_\mu(u,\bar{v}) = \frac{(w_\mu)_u(u)\overline{(w_\mu)_v(v)}}{(w_\mu(u) - \overline{w_\mu(v)})^2}.$$

We have from (1.21) the following formulas [**Ahl62**]

(1.24)
$$\frac{\partial}{\partial \varepsilon} K_{\varepsilon\mu}(z,u) = -\frac{1}{\pi}\iint_{\mathbb{U}} \mu(v) K_{\varepsilon\mu}(z,v) K_{\varepsilon\mu}(v,u) \, d^2v,$$

$$\frac{\partial}{\partial \bar{\varepsilon}} K_{\varepsilon\mu}(z,u) = -\frac{1}{\pi}\iint_{\mathbb{U}} \overline{\mu(v)} K_{\varepsilon\mu}(z,\bar{v}) K_{\varepsilon\mu}(\bar{v},u) \, d^2v,$$

and

$$\text{(1.25)} \qquad \frac{\partial}{\partial \varepsilon} K_{\varepsilon\mu}(z, \bar{u}) = -\frac{1}{\pi} \iint_{\mathbb{U}} \mu(v) K_{\varepsilon\mu}(z, v) K_{\varepsilon\mu}(v, \bar{u}) \, d^2v,$$

$$\frac{\partial}{\partial \bar{\varepsilon}} K_{\varepsilon\mu}(z, \bar{u}) = -\frac{1}{\pi} \iint_{\mathbb{U}} \overline{\mu(v)} K_{\varepsilon\mu}(z, \bar{v}) K_{\varepsilon\mu}(\bar{v}, \bar{u}) \, d^2v,$$

where the integrals are understood in the principal value sense.

For the model $\mathbb{H}^2 \simeq \mathbb{D}$ the q.c. mapping w_μ is normalized by fixing $-1, -i, 1$. The kernel R is given by

$$R(z, u) = \frac{(z+1)(z+i)(z-1)}{(u-z)(u+1)(u+i)(u-1)},$$

and formulas similar to (1.22) hold for F and Φ. In particular, let f be a q.c. mapping such that $f|_{\mathbb{D}} \in \mathcal{D}$, and let μ be a Beltrami differential supported on the quasi-disk $\Omega^* = f(\mathbb{D}^*)$. Let $v_{t\mu}$ be the solution on $\mathbb{C}$ of the Beltrami equation

$$(v_{t\mu})_{\bar{z}} = t\mu (v_{t\mu})_z,$$

satisfying $v_{t\mu}(0) = 0, v'_{t\mu}(0) = 1$ and $v''_{t\mu}(0) = 0$. Then

$$\dot{v} = \left.\frac{d}{dt}\right|_{t=0} v_{t\mu}$$

is a holomorphic function on $\Omega = f(\mathbb{D})$ and

$$\text{(1.26)} \qquad \dot{v}_{zzz}(z) = -\frac{6}{\pi} \iint_{\Omega^*} \frac{\mu(u)}{(u-z)^4} d^2u.$$

2. $T(1)$ as a Hilbert manifold

In this section we are going to endow $T(1)$ with a structure of a complex manifold modeled on the separable Hilbert space

$$A_2(\mathbb{D}) = \left\{ \phi \text{ holomorphic on } \mathbb{D} : \|\phi\|_2^2 = \iint_{\mathbb{D}} |\phi|^2 \rho^{-1}(z) d^2z < \infty \right\}$$

of holomorphic functions on $\mathbb{D}$. In the corresponding topology, the universal Teichmüller space $T(1)$ is a disjoint union of uncountably many components on which the right translations act transitively.

2.1. Hilbert space structure on tangent spaces. Let

$$A_2(\mathbb{D}^*) = \left\{ \phi \text{ holomorphic on } \mathbb{D}^* : \|\phi\|_2^2 = \iint_{\mathbb{D}^*} |\phi|^2 \rho^{-1}(z) d^2z < \infty \right\}$$

be the Hilbert space of holomorphic functions on $\mathbb{D}^*$.

LEMMA 2.1. *The vector spaces $A_2(\mathbb{D})$ and $A_2(\mathbb{D}^*)$ are subspaces of $A_\infty(\mathbb{D})$ and $A_\infty(\mathbb{D}^*)$ respectively. The natural inclusion maps $A_2(\mathbb{D}) \hookrightarrow A_\infty(\mathbb{D})$ and $A_2(\mathbb{D}^*) \hookrightarrow A_\infty(\mathbb{D}^*)$ are bounded linear mappings of Banach spaces.*

PROOF. It is sufficient to consider only the spaces of holomorphic functions on $\mathbb{D}$. For every $\phi \in A_2(\mathbb{D})$, let $\phi = \sum_{n=2}^{\infty}(n^3-n)a_n z^{n-2}$ be the power series expansion. Then
$$\|\phi\|_2^2 = \iint_{\mathbb{D}} |\phi|^2 \rho^{-1} d^2 z = \frac{\pi}{2}\sum_{n=2}^{\infty}(n^3-n)|a_n|^2,$$
and by Cauchy-Schwarz inequality,
$$|\phi(z)| = \left|\sum_{n=2}^{\infty}(n^3-n)a_n z^{n-2}\right|$$
$$\leq \left(\sum_{n=2}^{\infty}(n^3-n)|a_n|^2\right)^{1/2} \left(\sum_{n=2}^{\infty}(n^3-n)|z|^{2n-4}\right)^{1/2}$$
for every $z \in \mathbb{D}$. Since
$$\sum_{n=2}^{\infty}(n^3-n)|z|^{2n-4} = \frac{6}{(1-|z|^2)^4},$$
we have
$$\|\phi\|_{\infty} = \sup_{z\in\mathbb{D}}\left|(1-|z|^2)^2 \phi(z)\right| \leq \sqrt{\frac{12}{\pi}}\|\phi\|_2.$$
$\square$

Let
$$H^{-1,1}(\mathbb{D}) = \left\{\mu = \rho^{-1}\bar{\phi},\ \phi \text{ holomorphic on } \mathbb{D} : \|\mu\|_2^2 = \iint_{\mathbb{D}}|\mu|^2 \rho(z) d^2 z < \infty\right\}$$
and
$$H^{-1,1}(\mathbb{D}^*) = \left\{\mu = \rho^{-1}\bar{\phi},\ \phi \text{ holomorphic on } \mathbb{D}^* : \|\mu\|_2^2 = \iint_{\mathbb{D}^*}|\mu|^2 \rho(z) d^2 z < \infty\right\}$$
be the Hilbert spaces of harmonic Beltrami differentials on $\mathbb{D}$ and $\mathbb{D}^*$ respectively. It follows from Lemma 2.1 that the natural inclusion maps $H^{-1,1}(\mathbb{D}) \hookrightarrow \Omega^{-1,1}(\mathbb{D})$ and $H^{-1,1}(\mathbb{D}^*) \hookrightarrow \Omega^{-1,1}(\mathbb{D}^*)$ are bounded and under the linear mapping $D_0\beta$, $H^{-1,1}(\mathbb{D}^*) \xrightarrow{\sim} A_2(\mathbb{D})$.

REMARK 2.2. It follows from the proof of Lemma 2.1 that every $\mu \in H^{-1,1}(\mathbb{D})$ (respectively in $H^{-1,1}(\mathbb{D}^*)$) satisfies
$$\lim_{|z|\to 1}\mu(z) = 0.$$
Indeed, for given $\varepsilon > 0$ let N be such that
$$\sum_{n=N}^{\infty}(n^3-n)|a_n|^2 < \varepsilon.$$

Then
$$|\mu(z)| \leq \frac{(1-|z|^2)^2}{4} \left|\sum_{n=2}^{N-1}(n^3-n)a_n z^{n-2}\right|$$
$$+ \frac{(1-|z|^2)^2}{4}\left(\sum_{n=N}^{\infty}(n^3-n)|a_n|^2\right)^{1/2}\left(\sum_{n=2}^{\infty}(n^3-n)|z|^{2n-4}\right)^{1/2}$$
$$\leq \frac{(1-|z|^2)^2}{4}\left|\sum_{n=2}^{N-1}(n^3-n)a_n z^{n-2}\right| + \frac{\sqrt{6}\varepsilon}{4},$$

so that
$$\limsup_{|z|\to 1}|\mu(z)| \leq \frac{\sqrt{6}\varepsilon}{4}.$$

Since ε is arbitrary this proves the assertion.

For every $[\mu] \in T(1)$ let $D_0 R_{[\mu]}\left(H^{-1,1}(\mathbb{D}^*)\right)$ be the subspace of the tangent space $T_{[\mu]}T(1) = D_0 R_{[\mu]}\left(\Omega^{-1,1}(\mathbb{D}^*)\right)$ with a Hilbert space structure isomorphic to $H^{-1,1}(\mathbb{D}^*)$. Let $\mathfrak{D}_T$ be the distribution on $T(1)$, defined by the assignment
$$T(1) \ni [\mu] \mapsto D_0 R_{[\mu]}\left(H^{-1,1}(\mathbb{D}^*)\right) \subset T_{[\mu]}T(1).$$
Similarly, let $\mathfrak{D}_A$ be the distribution on $A_\infty(\mathbb{D})$, defined by
$$A_\infty(\mathbb{D}) \ni \phi \mapsto A_2(\mathbb{D}) \subset T_\phi A_\infty(\mathbb{D}) \simeq A_\infty(\mathbb{D}).$$

The next statement asserts that under the Bers embedding $\beta : T(1) \to A_\infty(\mathbb{D})$ the distribution $\mathfrak{D}_T$ is isomorphic to the restriction of the distribution $\mathfrak{D}_A$ to $\beta(T(1))$.

THEOREM 2.3. *For every $[\mu] \in T(1)$ the linear mapping*
$$D_0\left(\beta \circ R_{[\mu]}\right) : H^{-1,1}(\mathbb{D}^*) \to A_2(\mathbb{D})$$
is a topological isomorphism.

PROOF. Let $\nu \in H^{-1,1}(\mathbb{D}^*)$. Set $w_t = w_{t\nu*\mu} = w_{t\nu} \circ w_\mu$ and let $w_t = g_t^{-1} \circ f^t$ be the conformal welding associated with the q.c. mapping w_t by (1.15). Let $v_t = f^t \circ f^{-1}$, where $w_\mu = g^{-1} \circ f$ is the factorization for w_μ, and set $\Omega = f(\mathbb{D}) = g(\mathbb{D})$, $\Omega^* = f(\mathbb{D}^*) = g(\mathbb{D}^*)$. Since
$$\beta([t\nu*\mu]) = \mathcal{S}(f^t) = \mathcal{S}(v_t) \circ f \, f_z^2 + \mathcal{S}(f),$$
we have
$$D_0\left(\beta \circ R_{[\mu]}\right)(\nu) = \left.\frac{d}{dt}\right|_{t=0} \mathcal{S}(f^t) = \dot{v}_{zzz} \circ f \, f_z^2, \quad \text{where} \quad \dot{v} = \left.\frac{d}{dt}\right|_{t=0} v_t.$$
The q.c. mapping v_t is holomorphic on Ω and satisfies $v_t \circ g = g_t \circ w_{t\nu}$. Since g_t and g are holomorphic on $\mathbb{D}^*$, the Beltrami differential of v_t is given by
$$t\tilde{\nu}(z) = \begin{cases} 0, & z \in \Omega, \\ t(\nu \circ g^{-1})(z)\overline{\dfrac{g_z^{-1}(z)}{g_z^{-1}(z)}}, & z \in \Omega^*. \end{cases}$$

2. $T(1)$ AS A HILBERT MANIFOLD

It follows from (1.26) that

$$(2.1) \quad D_0\left(\beta \circ R_{[\mu]}\right)(\nu)\left(\mathrm{f}^{-1}(z)\right)\left(\mathrm{f}_z^{-1}(z)\right)^2 = \dot{v}_{zzz}(z) = -\frac{6}{\pi}\iint_{\Omega^*}\frac{\tilde{\nu}(u)}{(u-z)^4}d^2u.$$

Let $\rho_1(z) = (\rho \circ \mathrm{f}^{-1})(z)|\mathrm{f}_z^{-1}(z)|^2$ and $\rho_2(z) = (\rho \circ \mathrm{g}^{-1})(z)|\mathrm{g}_z^{-1}(z)|^2$ be the hyperbolic metric densities on the domains Ω and Ω^* respectively. Classical inequalities (see e.g., [**Leh87**, **Nag88**])

$$\frac{1}{4} \leq \eta_i^2(z)\rho_i(z) \leq 4, \qquad i=1,2,$$

where $\eta_1(z)$ and $\eta_2(z)$ stand, respectively, for the distances of $z \in \Omega$ and $z \in \Omega^*$ to the quasi-circle $\mathrm{f}(S^1)$, yield the following estimates (cf. [**Nag88**, Sect. 3.4.5])

$$\iint_{\Omega}\frac{d^2z}{|u-z|^4} \leq \iint_{|z-u|\geq\eta_2(u)}\frac{d^2z}{|u-z|^4} = \frac{\pi}{\eta_2(u)^2} \leq 4\pi\rho_2(u), \quad u \in \Omega^*,$$

and

$$\iint_{\Omega^*}\frac{d^2u}{|u-z|^4} \leq 4\pi\rho_1(z), \quad z \in \Omega.$$

From here it follows

$$\|D_0\left(\beta \circ R_{[\mu]}\right)(\nu)\|_2^2 = \iint_{\mathbb{D}}\left|D_0\left(\beta \circ R_{[\mu]}\right)(\nu)\right|^2\rho^{-1}d^2z = \iint_{\Omega}|\dot{v}_{zzz}|^2\rho_1^{-1}d^2z$$

$$\leq \frac{6^2}{\pi^2}\iint_{\Omega}\rho_1(z)^{-1}\iint_{\Omega^*}\frac{d^2v}{|v-z|^4}\iint_{\Omega^*}\frac{|\tilde{\nu}(u)|^2d^2u}{|u-z|^4}\,d^2z$$

$$\leq \frac{6^2 \cdot 4}{\pi}\iint_{\Omega}\iint_{\Omega^*}\frac{|\tilde{\nu}(u)|^2d^2u}{|u-z|^4}\,d^2z \leq 6^2 \cdot 4^2\iint_{\Omega^*}|\tilde{\nu}(u)|^2\rho_2(u)d^2u$$

$$= 6^2 \cdot 4^2 \iint_{\mathbb{D}^*}|\nu|^2\rho(u)d^2u = 576\|\nu\|_2^2.$$

To prove that the mapping $D_0\left(\beta \circ R_{[\mu]}\right)$ is onto, we adapt to our case Bers' arguments, as presented in [**Nag88**, Sect. 3.5]. For $\phi \in A_2(\mathbb{D})$ set $q = (\phi \circ \mathrm{f}^{-1})(\mathrm{f}_z^{-1})^2$, and choose μ in the equivalence class of $[\mu] \in T(1)$ to be the conformally natural extension of $(\mathrm{g}^{-1} \circ \mathrm{f})|_{S^1}$, constructed by Douady and Earle [**DE86**]. Let h be the corresponding quasiconformal reflection [**EN88**] on $\mathbb{C}$ which fixes the quasi-circle $\mathrm{f}(S^1)$. According to the Bers reproducing formula [**Ber66**],

$$(2.2) \quad q(z) = -\frac{3}{\pi}\iint_{\Omega^*}\frac{q(h(u))(u-h(u))^2 h_{\bar{u}}(u)}{(u-z)^4}d^2u.$$

Analogous to $L^\infty(\mathbb{D}^*)$ and $\Omega^{-1,1}(\mathbb{D}^*)$, consider the Banach spaces $L^\infty(\Omega^*)$ and

$$\Omega^{-1,1}(\Omega^*) = \left\{\mu \in L^\infty(\Omega^*) : \mu = \rho_2^{-1}\bar{q},\ q \text{ is holomorphic on } \Omega^*\right\}.$$

Denote by $\tilde{P}$ the corresponding projection $\tilde{P} : L^\infty(\Omega^*) \to \Omega^{-1,1}(\Omega^*)$. The mapping

$$\mu \mapsto (\mathrm{g}^*)^{-1}(\mu) = \mu \circ \mathrm{g}^{-1}\frac{\overline{(\mathrm{g}^{-1})'}}{(\mathrm{g}^{-1})'}$$

establishes the isomorphisms $L^\infty(\mathbb{D}^*) \simeq L^\infty(\Omega^*)$ and $\Omega^{-1,1}(\mathbb{D}^*) \simeq \Omega^{-1,1}(\Omega^*)$, and $\tilde{P} = (g^*)^{-1} \circ P \circ g^*$. Define $\nu \in \Omega^{-1,1}(\mathbb{D}^*)$ by

$$(g^*)^{-1}(\nu)(z) = \tilde{P}\left(\frac{1}{2}(q(h(z))(z-h(z))^2 h_{\bar{z}}(z)\right) \in \Omega^{-1,1}(\Omega^*).$$

The comparison between (2.2) and (2.1) shows that

$$D_0\left(\beta \circ R_{[\mu]}\right)(\nu) = \phi.$$

To prove that $\nu \in H^{-1,1}(\mathbb{D}^*)$ we use the Earle-Nag [**EN88**] estimate,

(2.3) $$\frac{1}{C} \leq |z - h(z)|^4 \rho_1(h(z))\rho_2(z) \leq C, \quad z \in \Omega^*,$$

where the constant C depends only on $\|\mu\|_\infty$. Since the operator $\tilde{P}$ gives the orthogonal projection of $L^2(\Omega^*, \rho_2(z)\,d^2z)$ onto $H^{-1,1}(\Omega^*)$, we get by the Earle-Nag inequality

$$\iint_{\mathbb{D}^*} |\nu|^2 \rho(z) d^2z = \iint_{\Omega^*} |(g^*)^{-1}(\nu)(z)|^2 \rho_2(z) d^2z$$

$$\leq \frac{1}{4} \iint_{\Omega^*} |q(h(z))(z-h(z))^2 h_{\bar{z}}(z)|^2 \rho_2(z) d^2z$$

$$\leq \frac{C}{4} \iint_{\Omega^*} |q(h(z)) h_{\bar{z}}(z)|^2 \rho_1^{-1}(h(z)) d^2z.$$

Since h is sense reversing, for

$$\kappa = \frac{h_z^{-1}}{h_{\bar{z}}^{-1}}$$

we have $\|\kappa\|_\infty < 1$. Now

$$\iint_{\Omega^*} |q(h(z)) h_{\bar{z}}(z)|^2 \rho_1(h(z))^{-1} d^2z$$

$$= \iint_\Omega |q(z)|^2 \rho_1(z)^{-1} |(h_{\bar{z}} \circ h^{-1})(z) h_{\bar{z}}^{-1}(z)|^2 (1 - |\kappa(z)|^2) d^2z$$

$$= \iint_\Omega \frac{|q(z)|^2}{1 - |\kappa(z)|^2} \rho_1(z)^{-1} d^2z$$

$$\leq C_1 \iint_\Omega |q(z)|^2 \rho_1(z)^{-1} d^2z = C_1 \|\phi\|_2^2 < \infty,$$

so that $\|\nu\|_2 \leq C_2 \|\phi\|_2$. This also proves that the inverse map to $D_0\left(\beta \circ R_{[\mu]}\right)$ is bounded, so that $D_0\left(\beta \circ R_{[\mu]}\right)$ is a topological isomorphism. $\square$

REMARK 2.4. It follows from the proof of the first part of Theorem 2.3 that the operator $D_0\left(\beta \circ R_{[\mu]}\right) = D_0\left(\beta \circ \Phi \circ R_\mu\right)$ extends to the bounded linear operator on $L^2(\mathbb{D}^*, \rho(z) d^2z)$ and the estimate

$$\|D_0\left(\beta \circ \Phi \circ R_\mu\right)(\nu)\|_2 = \|D_\mu\left(\beta \circ \Phi\right)(D_0 R_\mu(\nu))\|_2 \leq 24\|\nu\|_2$$

holds for all $\nu \in L^2(\mathbb{D}^*, \rho(z)d^2z)$ and $\mu \in L^\infty(\mathbb{D}^*)_1$.

2.2. The L^2-estimates.
The lemmas below are needed for the rigorous definition of a complex Hilbert manifold structure on $T(1)$.

LEMMA 2.5. *For every $\varepsilon > 0$ there exists $0 < \delta < 1$ such that for all $\mu \in \Omega^{-1,1}(\mathbb{D}^*)$ with $\|\mu\|_\infty < \delta$,*

$$\left| \frac{|(w_\mu)_z(z)|^2}{(1-|w_\mu(z)|^2)^2} - \frac{1}{(1-|z|^2)^2} \right| < \frac{\varepsilon}{(1-|z|^2)^2}$$

for all $z \in \mathbb{D} \cup \mathbb{D}^*$. The same inequality holds for $w_{\mu^{-1}} = w_\mu^{-1}$.

PROOF. Using the isomorphism $\Omega^{-1,1}(\mathbb{D}^*) \xrightarrow{\sim} \Omega^{-1,1}(\mathbb{D})$ given by reflection (1.1) and property (1.2), it is sufficient to prove the estimate for $z \in \mathbb{D}$. Since $g_\mu = \alpha \circ w_\mu$, where $\alpha \in \mathrm{PSU}(1,1)$, the estimate holds for w_μ if and only if it holds for g_μ. By Lemma 1.5 g_μ fixes 0 and ∞, and by the result of Ahlfors and Bers in [**AB60**] (see also the remark of Bers in [**Ber73**]) the functional

$$\mathcal{L}^\infty(\mathbb{D}^*)_1 \ni \mu \mapsto (g_\mu)_z(0) \in \mathbb{C}$$

is real-analytic at $\mu = 0$. In particular, for every $\varepsilon > 0$ there exists $0 < \delta < 1$ such that for all $\mu \in \Omega^{-1,1}(\mathbb{D}^*)_1$ with $\|\mu\|_\infty < \delta$,

$$\left| |(g_\mu)_z(0)|^2 - 1 \right| < \varepsilon.$$

For $z \in \mathbb{D}$, let $\tilde{\mu} = \mu \circ \sigma_z$, where $\sigma_z(w) = \frac{w+z}{1+\bar{z}w}$. Then $g_{\tilde{\mu}} = \tilde{\sigma}_z \circ g_\mu \circ \sigma_z$ for some $\tilde{\sigma}_z \in \mathrm{PSU}(1,1)$. Since $\tilde{\mu} \in \Omega^{-1,1}(\mathbb{D}^*)_1$, it follows from Lemma 1.5 that $g_{\tilde{\mu}}(0) = 0$. Therefore from $\tilde{\sigma}_z(g_\mu(z)) = 0$, one obtains

$$\frac{|(g_\mu)_z(z)|^2}{(1-|g_\mu(z)|^2)^2}(1-|z|^2)^2 = |(g_{\tilde{\mu}})_z(0)|^2.$$

Since $\|\tilde{\mu}\|_\infty = \|\mu\|_\infty$, the assertion follows. On the other hand, if $\mu \in \Omega^{-1,1}(\mathbb{D}^*)_1$, then $\mu^{-1} \in \mathcal{L}^\infty(\mathbb{D}^*)_1$ and $\|\mu^{-1}\|_\infty = \|\mu\|_\infty$. Hence the assertion also holds for $w_{\mu^{-1}}$. □

COROLLARY 2.6. *Let $\mu \in \Omega^{-1,1}(\mathbb{D}^*)$, $\|\mu\|_\infty < \delta$, where δ corresponds to $\varepsilon = 1$ in the previous lemma. Then for every $\lambda \in L^\infty(\mathbb{D}^*)_1$ the linear mapping $D_\lambda R_\mu$ extends to an invertible bounded linear operator on the Hilbert space $L^2(\mathbb{D}^*, \rho(z)d^2z)$. Moreover,*

$$\left\| D_\lambda R_\mu(\nu) \right\|_2 \leq \frac{\sqrt{2}}{(1-\|\mu\|_\infty)^2} \|\nu\|_2,$$

for all $\nu \in L^2(\mathbb{D}^*, \rho(z)d^2z)$ and $\lambda \in L^\infty(\mathbb{D}^*)_1$. The same inequality holds for $D_\lambda R_{\mu^{-1}}$.

PROOF. Since

$$D_\lambda R_\mu(\nu) = \frac{(1-|\mu|^2)\,\nu \circ w_\mu}{\left(1 + \bar{\mu}\lambda \circ w_\mu \frac{\overline{(w_\mu)_z}}{(w_\mu)_z}\right)^2} \frac{\overline{(w_\mu)_z}}{(w_\mu)_z},$$

and

$$\left\| \frac{1-|\mu|^2}{\left(1 + \bar{\mu}\lambda \circ w_\mu \frac{\overline{(w_\mu)_z}}{(w_\mu)_z}\right)^2} \right\|_\infty \leq \frac{1}{(1-\|\mu\|_\infty)^2}$$

for all $\lambda \in L^\infty(\mathbb{D}^*)_1$, we have by using Lemma 2.5 and $\|\mu\|_\infty = \|\mu^{-1}\|_\infty$,

$$\iint_{\mathbb{D}^*} \left|D_\lambda R_\mu(\nu)\right|^2 \rho(z) d^2z \leq (1-\|\mu\|_\infty)^{-4} \iint_{\mathbb{D}^*} \left|\nu \circ w_\mu \frac{\overline{(w_\mu)_z}}{(w_\mu)_z}\right|^2 \rho(z) d^2z$$

$$= (1-\|\mu\|_\infty)^{-4} \iint_{\mathbb{D}^*} |\nu|^2 \frac{4|(w_{\mu^{-1}})_z|^2}{(1-|w_{\mu^{-1}}|^2)^2}(1-|\mu^{-1}|^2) d^2z$$

$$\leq 2(1-\|\mu\|_\infty)^{-4} \iint_{\mathbb{D}^*} |\nu|^2 \rho(z) d^2z = 2(1-\|\mu\|_\infty)^{-4} \|\nu\|_2^2.$$

Replacing everywhere μ by μ^{-1} we get the same estimate for $D_\lambda R_{\mu^{-1}}$. $\square$

Denote by $\mathcal{O}(\mathbb{D}^*)_1$ the subgroup of $L^\infty(\mathbb{D}^*)_1$ generated by $\mu \in \Omega^{-1,1}(\mathbb{D}^*)$, $\|\mu\|_\infty < \delta$, where δ is as in Corollary 2.6.

LEMMA 2.7. *For every $\mu \in \mathcal{O}(\mathbb{D}^*)_1$ there exists $C > 0$ such that*

$$\|R_\mu(\lambda_1) - R_\mu(\lambda_2)\|_2 < C\|\lambda_1 - \lambda_2\|_2$$

for all $\lambda_1, \lambda_2 \in L^\infty(\mathbb{D}^*)_1$ satisfying $\lambda_1 - \lambda_2 \in L^2(\mathbb{D}^*, \rho(z) d^2z)$.

PROOF. Suppose first that $\|\mu\|_\infty < \delta$. Set $\lambda(t) = \lambda_1 + t\nu$, where $\nu = \lambda_2 - \lambda_1$, so that $\lambda(t) \in L^\infty(\mathbb{D}^*)_1$, $0 \leq t \leq 1$. By fundamental theorem of calculus,

$$R_\mu(\lambda_1) - R_\mu(\lambda_2) = \int_0^1 \frac{d}{dt} R_\mu(\lambda(t)) dt$$

$$= \int_0^1 D_{\lambda(t)} R_\mu(\nu) dt.$$

Using Corollary 2.6,

$$\|R_\mu(\lambda_1) - R_\mu(\lambda_2)\|_2^2 = \iint_{\mathbb{D}^*} \left|\int_0^1 D_{\lambda(t)} R_\mu(\nu)(z)\right|^2 \rho(z) d^2z$$

$$\leq \int_0^1 \left(\iint_{\mathbb{D}^*} \left|D_{\lambda(t)} R_\mu(\nu)(z)\right|^2 \rho(z) d^2z\right) dt$$

$$\leq C^2 \|\nu\|_2^2 = C^2 \|\lambda_1 - \lambda_2\|_2^2.$$

The same estimate also holds for R_μ^{-1}.

Since every $\mu \in \mathcal{O}(\mathbb{D}^*)_1$ can be written as $\mu_n^{\varepsilon_n} * \cdots * \mu_1^{\varepsilon_1}$, where $\mu_i \in \Omega^{-1,1}(\mathbb{D}^*)$, $\|\mu_i\|_\infty < \delta$, and $\varepsilon_i = \pm 1$, $i = 1, \ldots, n$, we have

$$R_\mu = R_{\mu_1}^{\varepsilon_1} \circ \cdots \circ R_{\mu_n}^{\varepsilon_n},$$

and the assertion of the lemma follows. $\square$

REMARK 2.8. Applying the same argument, we get from Corollary 2.6 that for every $\mu \in \mathcal{O}(\mathbb{D}^*)_1$ there exists $C > 0$, depending only on $\|\mu\|_\infty$ such that

$$\left\|D_\lambda R_\mu(\nu)\right\|_2 \leq C\|\nu\|_2,$$

for all $\nu \in L^2(\mathbb{D}^*, \rho(z) d^2z)$ and $\lambda \in L^\infty(\mathbb{D}^*)_1$.

LEMMA 2.9. *For every $\mu \in \mathcal{O}(\mathbb{D}^*)_1$ there exists $C > 0$ such that*
$$\|(\beta \circ \Phi)(\lambda * \mu) - (\beta \circ \Phi)(\mu)\|_2 \leq C\|\lambda\|_2$$
for all $\lambda \in L^2(\mathbb{D}^*, \rho(z)d^2z) \cap L^\infty(\mathbb{D}^*)_1$.

PROOF. Set $\phi(t) = (\beta \circ \Phi)(t\lambda * \mu)$. By fundamental theorem of calculus,
$$(\beta \circ \Phi)(\lambda * \mu) - (\beta \circ \Phi)(\mu) = \int_0^1 \frac{d\phi}{dt}(t)dt,$$
where
$$\frac{d\phi}{dt}(t) = D_{t\lambda}\left(\beta \circ \Phi \circ R_\mu\right)(\lambda) = \left(D_{t\lambda*\mu}(\beta \circ \Phi) \circ D_{t\lambda}R_\mu\right)(\lambda)$$
by chain rule. Since $(D_0 R_\mu)^{-1} = D_\mu R_{\mu^{-1}}$, it follows from Remarks 2.4 and 2.8 that
$$\|D_{t\lambda*\mu}(\beta \circ \Phi)(\nu)\|_2 \leq 24\|D_{t\lambda*\mu}R_{(t\lambda*\mu)^{-1}}(\nu)\|_2 \leq C_1\|\nu\|_2.$$
Using Remark 2.8 again, we get
$$\left\|\frac{d\phi}{dt}(t)\right\|_2 = \|\left(D_{t\lambda*\mu}(\beta \circ \Phi) \circ D_{t\lambda}R_\mu\right)(\lambda)\|_2 \leq C_2\|\lambda\|_2, \quad 0 \leq t \leq 1.$$
Therefore,
$$\|(\beta \circ \Phi)(\lambda*\mu) - (\beta \circ \Phi)(\mu)\|_2^2 = \iint_\mathbb{D} \left|\int_0^1 \frac{d\phi}{dt}(t,z)dt\right|^2 \rho(z)^{-1}d^2z$$
$$\leq \int_0^1 \left(\iint_\mathbb{D} \left|\frac{d\phi}{dt}(t,z)\right|^2 \rho(z)^{-1}d^2z\right)dt$$
$$\leq C_2^2 \|\lambda\|_2^2,$$
which concludes the proof. $\square$

2.3. The Hilbert manifold structure of $T(1)$. For every $\mu \in \mathcal{O}(\mathbb{D}^*)_1$ let $V_\mu \subset U_\mu \subset T(1)$ be the image under the map $h_\mu^{-1} = \Phi \circ R_\mu \circ \Lambda$ of the open ball of radius $\sqrt{\pi/3}$ about the origin in $A_2(\mathbb{D})$, which by Lemma 2.1 is contained in the ball of radius 2 in $A_\infty(\mathbb{D})$. Here (U_μ, h_μ) is the coordinate chart U_μ of the complex-analytic atlas for $T(1)$ as a complex Banach manifold (see Section 1.1.4). Let
$$\tilde{h}_\mu = h_\mu|_{V_\mu} : V_\mu \to A_2(\mathbb{D}).$$
The main result of this subsection is the following.

THEOREM 2.10. *For every $\mu, \nu \in \mathcal{O}(\mathbb{D}^*)_1$ the sets $\tilde{h}_\mu(V_\mu \cap V_\nu)$ and $\tilde{h}_\nu(V_\mu \cap V_\nu)$ are open in $A_2(\mathbb{D})$ and the map*
$$\tilde{h}_{\mu\nu} = \tilde{h}_\mu \circ \tilde{h}_\nu^{-1} : \tilde{h}_\nu(V_\mu \cap V_\nu) \longrightarrow \tilde{h}_\mu(V_\mu \cap V_\nu) \subset A_2(\mathbb{D})$$
is a biholomorphic function on the Hilbert space $A_2(\mathbb{D})$.

PROOF. First we prove that the sets $\tilde{h}_\mu(V_\mu \cap V_\nu)$ and $\tilde{h}_\nu(V_\mu \cap V_\nu)$ are open in $A_2(\mathbb{D})$. Since $V_\mu \cap V_\nu \neq \emptyset$ (otherwise there is nothing to prove), there exist $\phi_1 \in \tilde{h}_\mu(V_\mu \cap V_\nu)$ and $\phi_2 \in \tilde{h}_\nu(V_\mu \cap V_\nu)$, $\|\phi_1\|_2, \|\phi_2\|_2 < \sqrt{\pi/3}$, such that $\tilde{h}_\mu^{-1}(\phi_1) = \tilde{h}_\nu^{-1}(\phi_2)$, i.e.,
$$(\Phi \circ R_\mu \circ \Lambda)(\phi_1) = (\Phi \circ R_\nu \circ \Lambda)(\phi_2).$$

Setting $\lambda_1 = \Lambda(\phi_1)$, $\lambda_2 = \Lambda(\phi_2)$ and $\kappa = \nu * \mu^{-1}$, we get
$$\Phi(\lambda_1) = \Phi(\lambda_2 * \kappa).$$

The sets $h_\mu(U_\mu \cap U_\nu)$ and $h_\nu(U_\mu \cap U_\nu)$ are open in $A_\infty(\mathbb{D})$, so that there exists $\delta_1 > 0$ such that $h_\mu(U_\mu \cap U_\nu)$ contains a ball of radius δ_1 about ϕ_1 in $A_\infty(\mathbb{D})$. The mapping $h_{\mu\nu} : h_\nu(U_\mu \cap U_\nu) \to h_\mu(U_\mu \cap U_\nu) \subset A_\infty(\mathbb{D})$ is a continuous function in the Banach space $A_\infty(\mathbb{D})$, so that there exists $\delta_2 > 0$ such that the inverse image by $h_{\mu\nu}$ of the ball of radius δ_1 about ϕ_1 in $A_\infty(\mathbb{D})$ contains the ball of radius δ_2 about ϕ_2 in $A_\infty(\mathbb{D})$. According to Lemma 2.1, the latter ball contains any ball of radius $\delta_3 < \sqrt{\pi/12}\,\delta_2$ about ϕ_2 in $A_2(\mathbb{D})$. Now for every $\varphi_2 \in A_2(\mathbb{D})$ satisfying $\|\varphi_2 - \phi_2\|_2 < \delta_3$ set
$$\varphi_1 = h_{\mu\nu}(\varphi_2) = (\beta \circ \Phi \circ R_\kappa \circ \Lambda)(\varphi_2).$$

We claim that $\delta_3 > 0$ can be chosen such that $\varphi_1 \in A_2(\mathbb{D})$ and $\|\varphi_1\|_2 < \sqrt{\pi/3}$, which implies that $\tilde{h}_\nu(V_\mu \cap V_\nu)$ contains the ball of radius δ_3 about ϕ_2 in $A_2(\mathbb{D})$. Indeed, set $\lambda = \Lambda(\varphi_2)$, so that $\varphi_1 = (\beta \circ \Phi)(\lambda * \kappa)$. Since $\lambda - \lambda_2 \in L^2(\mathbb{D}^*, \rho(z)d^2z)$, we have by Lemmas 2.9 and 2.7,
$$\|\varphi_1 - \phi_1\|_2 = \|(\beta \circ \Phi)(\lambda * \kappa) - (\beta \circ \Phi)(\lambda_2 * \kappa)\|_2$$
$$\leq C\|\lambda * \lambda_2^{-1}\|_2 \leq C^2\|\lambda - \lambda_2\|_2$$
$$= 2C^2\|\varphi_2 - \phi_2\|_2 < 2C^2\delta_3,$$

where the constant $C > 0$ (chosen to be the same for both Lemmas 2.7 and 2.9) depends only on λ_2 and κ. Choosing δ_3 small enough we have $\|\varphi_1\|_2 < \sqrt{\pi/3}$.

The same argument applied to the map $\tilde{h}_{\nu\mu} = \tilde{h}_{\mu\nu}^{-1}$ proves that $\tilde{h}_\mu(V_\mu \cap V_\nu)$ is open in $A_2(\mathbb{D})$.

It remains to prove that the map $\tilde{h}_{\mu\nu}$ is a holomorphic function in the Hilbert space $A_2(\mathbb{D})$. It is bounded, so according to [**Bou67**] it is sufficient to prove that for every $\varphi \in \tilde{h}_\nu(V_\mu \cap V_\nu)$ and every $\eta \in A_2(\mathbb{D})$ the mapping $\mathbb{C} \ni t \mapsto \phi(t) = \tilde{h}_{\mu\nu}(\varphi + t\eta) \in A_2(\mathbb{D})$ is a holomorphic function in some neighborhood of 0 in $\mathbb{C}$. For this purpose we use the standard argument based on the fact that the map $h_{\mu\nu}$ is already a holomorphic function in the Banach space $A_\infty(\mathbb{D})$ and the mapping $\mathbb{C} \ni t \mapsto \phi(t) \in A_\infty(\mathbb{D})$ is a holomorphic function in some neighborhood of 0 in $\mathbb{C}$. Thus there exists $\delta > 0$ such that for every $|t_0| < \delta$,
$$\left\|\phi(t) - \phi(t_0) - (t - t_0)\frac{d\phi}{dt}(t_0)\right\|_\infty = o(|t - t_0|) \quad \text{as} \quad t \to t_0.$$

Moreover, δ can be chosen such that $\varphi + t\eta \in \tilde{h}_\nu(V_\mu \cap V_\nu)$ for $|t| < \delta$. Then for every $z \in \mathbb{D}$ the complex-valued function $\phi(t)(z)$ is holomorphic on $|t| < \delta$ and
$$\phi(t,z) - \phi(t_0,z) - (t - t_0)\frac{d\phi}{dt}(t_0, z)$$
$$= \frac{1}{2\pi i} \oint_{|w-t_0|=\delta_1} \phi(w,z) \left(\frac{1}{w-t} - \frac{1}{w-t_0} - \frac{t-t_0}{(w-t_0)^2}\right) dw$$
$$= \frac{(t-t_0)^2}{2\pi i} \oint_{|w-t_0|=\delta_1} \frac{\phi(w,z)}{(w-t_0)^2(w-t)} dw,$$

where $\delta_1 > 0$ is such that the disk of radius δ_1 about t_0 is inside the disk of radius δ about the origin, and t satisfies $|t-t_0| < \delta_1$. Since $\phi(t) \in \tilde{h}_\mu(V_\mu \cap V_\nu)$, $\|\phi(t)\|_2^2 < \pi/3$

for all $|t| < \delta$, and we have

$$\left\|\frac{\phi(t) - \phi(t_0)}{t - t_0} - \frac{d\phi}{dt}(t_0)\right\|_2^2$$
$$\leq \frac{|t - t_0|^2}{4\pi^2} \oint_{|w|=\delta_1} \frac{|dw|}{|w|^4|w - (t-t_0)|^2} \oint_{|w|=\delta_1} \|\phi(w + t_0)\|_2^2 |dw|$$
$$= O(|t - t_0|^2) \quad \text{as} \quad t \to t_0.$$

$\square$

According to [**Bou67**], Theorem 2.10 justifies the following definition.

DEFINITION 2.11. The covering

$$T(1) = \bigcup_{\mu \in \mathcal{O}(\mathbb{D}^*)_1} V_\mu$$

with the coordinate maps $\tilde{h}_\mu : V_\mu \longrightarrow A_2(\mathbb{D})$ and the transition maps

$$\tilde{h}_{\mu\nu} = \tilde{h}_\mu \circ \tilde{h}_\nu^{-1} : \tilde{h}_\nu(V_\mu \cap V_\nu) \longrightarrow \tilde{h}_\mu(V_\mu \cap V_\nu)$$

is a complex-analytic atlas which endows $T(1)$ with the structure of a complex Hilbert manifold modeled on the Hilbert space $A_2(\mathbb{D})$.

COROLLARY 2.12. *The right translations are biholomorphic mappings on the Hilbert manifold $T(1)$.*

PROOF. Representing a point in $T(1)$ by $\mu \in \mathcal{O}(\mathbb{D}^*)_1$ we have $R_{[\mu]}(V_\lambda) = V_{\lambda*\mu}$, so that $\tilde{h}_{\lambda*\mu} \circ R_{[\mu]} \circ \tilde{h}_\lambda^{-1}$ is the identity mapping on $\tilde{h}_\lambda(V_\lambda) \subset A_2(\mathbb{D})$. $\square$

We will continue to use the name Bers coordinates for the complex coordinates $(V_\mu, \tilde{h}_\mu)$ on the Hilbert manifold $T(1)$. As in Section 1.1.4, the vector field $\frac{\partial}{\partial \varepsilon_\nu}$ corresponding to $\nu \in H^{-1,1}(\mathbb{D}^*)$ at a point $[\mu] \in V_0$ in terms of the Bers coordinates on V_μ has the same form (1.5), i.e.,

$$\frac{\partial}{\partial \varepsilon_\nu}\bigg|_{[\mu]} = P\left(\left(\frac{\nu}{1 - |\mu|^2}\frac{(w_\mu)_z}{(\overline{w}_\mu)_{\bar{z}}}\right) \circ w_\mu^{-1}\right),$$

where $P : L^2(\mathbb{D}^*, \rho(z)d^2z) \to H^{-1,1}(\mathbb{D}^*)$ is the orthogonal projector given by (1.7).

2.4. Integral manifolds of the distribution $\mathfrak{D}_T$. Finally, we introduce a Hilbert manifold structure on the Banach space $A_\infty(\mathbb{D})$ by defining the coordinate chart at every $\phi \in A_\infty(\mathbb{D})$ to be $\phi + A_2(\mathbb{D})$. By Lemma 2.1 the Hilbert manifold topology on $A_\infty(\mathbb{D})$ is stronger than the Banach space topology. The Hilbert manifold $A_\infty(\mathbb{D})$ is not connected. Rather $A_\infty(\mathbb{D})$ is the union of uncountably many components $\phi + A_2(\mathbb{D})$ with $\phi \in A_\infty(\mathbb{D})/A_2(\mathbb{D})$, which are integral manifolds of the distribution $\mathfrak{D}_A$.

THEOREM 2.13. *The Bers embedding $\beta : T(1) \to \beta(T(1)) \subset A_\infty(\mathbb{D})$ is a biholomorphic mapping of Hilbert manifolds.*

PROOF. To prove that the Bers embedding is holomorphic it is sufficient to show that for every $\mu \in \mathcal{O}(\mathbb{D}^*)_1$ the image of the ball of radius $\sqrt{\pi/3}$ about 0 in

$A_2(\mathbb{D})$ by the mapping $\beta \circ \tilde{h}_\mu^{-1} = \beta \circ \Phi \circ R_\mu \circ \Lambda$ is inside a translate by $(\beta \circ \Phi)(\mu)$ of some ball about 0 in $A_2(\mathbb{D})$. This immediately follows from Lemma 2.9,

$$\left\|\left(\beta \circ \tilde{h}_\mu^{-1}\right)(\varphi) - (\beta \circ \Phi)(\mu)\right\|_2 = \|(\beta \circ \Phi)(\lambda * \mu) - (\beta \circ \Phi)(\mu)\|_2$$
$$< C\|\lambda\|_2,$$

where $\lambda = \Lambda(\varphi) \in L^2(\mathbb{D}^*, \rho(z)d^2z) \cap L^\infty(\mathbb{D}^*)_1$ and the constant $C > 0$ depends only on $\|\mu\|_\infty$. Since the Bers embedding is a holomorphic mapping of Banach manifolds, the standard argument used in the proof of Theorem 2.10 works for this case, so that the mapping $\beta \circ \tilde{h}_\mu^{-1} - (\beta \circ \Phi)(\mu)$ is a holomorphic function on the Hilbert space $A_2(\mathbb{D})$.

Finally, the image $\beta(T(1))$ is open in the Hilbert manifold $A_\infty(\mathbb{D})$ since it is open in a weaker Banach manifold topology. Using Theorem 2.3 and the inverse function theorem for Hilbert manifolds [**Lan95**] we see that the Bers embedding is biholomorphic. □

Theorem 2.13 allows us to conclude that the distribution $\mathfrak{D}_T$ on $T(1)$ is equivalent to the restriction of the distribution $\mathfrak{D}_A$ on $\beta(T(1)) \subset A_\infty(\mathbb{D})$, and therefore is integrable. Its integral manifolds are inverse images by the Bers embedding β of the integral manifolds of the distribution $\mathfrak{D}_A$ on $\beta(T(1))$, i.e., of the components $(\phi + A_2(\mathbb{D})) \cap \beta(T(1))$. For every $[\mu] \in T(1)$ denote by $T_{[\mu]}(1)$ the component of the Hilbert manifold $T(1)$ containing $[\mu]$. It follows from Theorems 2.3 and 2.13 that the Hilbert manifold $T_{[\mu]}(1)$ is the integral manifold of the distribution $\mathfrak{D}_T$ passing through $[\mu] \in T(1)$. The right translations act transitively on the set of components, i.e. $R_{[\nu]}(T_{[\mu]}(1)) = T_{[\mu*\nu]}(1)$ for all $[\mu], [\nu] \in T(1)$.

The component of $0 \in T(1)$ by $T_0(1)$ is denoted by $T_0(1)$.

3. $T_0(1)$ as a topological group

Here we prove that the connected component $T_0(1)$ of $0 \in T(1)$ in the Hilbert manifold topology is the inverse image of $\beta(T(1)) \cap A_2(\mathbb{D})$ under the Bers embedding, and that $T_0(1)$ and $\mathcal{T}_0(1) = \pi^{-1}(T_0(1))$ are topological groups.

It follows from Theorem 2.13 that to prove $T_0(1) = \beta^{-1}(\beta(T(1)) \cap A_2(\mathbb{D}))$, it is sufficient to show that $\beta(T(1)) \cap A_2(\mathbb{D})$ is connected.

THEOREM 3.1. *The submanifold $\beta(T(1)) \cap A_2(\mathbb{D})$ of $\beta(T(1))$ is connected in the Hilbert manifold topology of $\beta(T(1))$.*

PROOF. First, in the Hilbert manifold topology the set $\beta(\text{Möb}(S^1)\backslash \text{Diff}_+(S^1))$ is dense in $\beta(T(1)) \cap A_2(\mathbb{D})$. Indeed, since $\beta(T(1)) \cap A_2(\mathbb{D})$ is open in $A_2(\mathbb{D})$, for every $\phi \in \beta(T(1)) \cap A_2(\mathbb{D})$ there exists δ such that $\tilde\phi \in \beta(T(1)) \cap A_2(\mathbb{D})$ for all $\tilde\phi$ satisfying $\|\tilde\phi - \phi\|_2 < \delta$. Since $\phi(z) = \sum_{k=2}^\infty (k^3 - k)a_k z^{k-2} \in A_2(\mathbb{D})$, for every $n \in \mathbb{N}$ there exists $N \in \mathbb{N}$ such that $\frac{\pi}{2}\sum_{k=N+1}^\infty (k^3 - k)|a_k|^2 < \frac{1}{n^2}$, so that for $\phi_n(z) = \sum_{k=2}^N (k^3 - k)a_k z^{k-2}$ we have $\|\phi_n - \phi\|_2 < \frac{1}{n}$. Thus for $\frac{1}{n} < \delta$ there exists $\gamma_n = g_n^{-1} \circ f_n \in \text{Möb}(S^1)\backslash \text{Homeo}_{qs}(S^1)$ such that $\mathcal{S}(f_n) = \phi_n$. Since ϕ_n is analytic on an open domain containing $\mathbb{D} \cup S^1$, so is the function f_n and, consequently, $f_n(S^1)$ is an analytic curve. Hence $\gamma_n|_{S^1}$ is smooth and $\phi_n \in \beta(\text{Möb}(S^1)\backslash \text{Diff}_+(S^1))$.

Secondly, the inclusion

$$\text{Möb}(S^1)\backslash \text{Diff}_+(S^1) \hookrightarrow T(1)$$

is a continuous mapping between the Frechet and Hilbert manifolds. Since the Frechet manifold $\mathrm{M\ddot{o}b}(S^1)\backslash\mathrm{Diff}_+(S^1)$ is connected, its image in the Hilbert manifold $T(1)$ is also connected. Finally, since Bers embedding is a continuous mapping (actually biholomorphic) and the closure of a connected set is connected, we conclude that $\beta(T(1)) \cap A_2(\mathbb{D})$ is connected in $\beta(T(1))$. $\square$

COROLLARY 3.2. *The Hilbert submanifold $T_0(1)$ of $T(1)$ is characterized by*
$$T_0(1) = \{[\mu] \in T(1) : \beta([\mu]) \in A_2(\mathbb{D})\}.$$

Next we prove that the Hilbert manifold $T_0(1)$ is a topological group.

LEMMA 3.3. *Every $[\mu] \in T_0(1)$ has a representative $\mu \in L^2(\mathbb{D}^*, \rho(z)d^2z) \cap \mathcal{O}(\mathbb{D}^*)_1$.*

PROOF. First observe that $L^2(\mathbb{D}^*, \rho(z)d^2z) \cap \mathcal{O}(\mathbb{D}^*)_1$ is a subgroup of $L^\infty(\mathbb{D}^*)_1$. Indeed, let $\lambda, \nu \in L^2(\mathbb{D}^*, \rho(z)d^2z) \cap \mathcal{O}(\mathbb{D}^*)_1$. Setting $\mu = \nu^{-1}$ and $\lambda_1 = \lambda, \lambda_2 = \nu$ in Lemma 2.7, we get
$$\|\lambda * \nu^{-1}\|_2 < C\|\lambda - \nu\|_2 < \infty.$$

Denote by B_α the ball of radius α about the origin in $A_2(\mathbb{D})$, where $\alpha < \sqrt{\frac{\pi}{3}}\delta$ and δ is defined in Corollary 2.6, and let $W_0 = (\Phi \circ \Lambda)(\beta(V_0) \cap B_\alpha)$. By definition, $W_0 \subset T_0(1)$ and every $[\mu] \in W_0$ has a representative $\mu \in L^2(\mathbb{D}^*, \rho(z)d^2z) \cap \mathcal{O}(\mathbb{D}^*)_1$. For every $\nu \in L^\infty(\mathbb{D}^*)_1$ set $W_\nu = R_{[\nu]}(W_0)$.

For every $[\mu] \in T_0(1)$ let $[0, 1] \ni t \mapsto [\mu(t)] \in T_0(1)$ be a path in $T_0(1)$ joining 0 to $[\mu]$. By definition of the Hilbert manifold structure on $T(1)$ (see Section 2.3), there exist $0 = t_0 < t_1 < \ldots < t_n = 1$ and $W_i = W_{\mu(t_i)} \subset T_0(1)$, which cover the path and such that $W_{i-1} \cap W_i \neq \emptyset$, $i = 1, \ldots, n$. We will prove by induction that each $[\mu(t_i)]$ has a representative $\mu_i \in L^2(\mathbb{D}^*, \rho(z)d^2z) \cap \mathcal{O}(\mathbb{D}^*)_1$. For $1 \leq i \leq n$, choose $[\nu_i] \in W_{i-1} \cap W_i$. Since $[\nu_1] \in W_0$, it has a representative $\nu_1 \in L^2(\mathbb{D}^*, \rho(z)d^2z) \cap \mathcal{O}(\mathbb{D}^*)_1$. On the other hand, since $[\nu_1] \in W_1$, $[\nu_1] = [\lambda * \mu(t_1)]$ for some $\lambda \in L^2(\mathbb{D}^*, \rho(z)d^2z) \cap \mathcal{O}(\mathbb{D}^*)_1$. Hence $[\mu(t_1)]$ has a representative $\mu_1 = \lambda^{-1} * \nu_1 \in L^2(\mathbb{D}^*, \rho(z)d^2z) \cap \mathcal{O}(\mathbb{D}^*)_1$. Now suppose that $[\mu(t_{i-1})]$ has a representative $\mu_{i-1} \in L^2(\mathbb{D}^*, \rho(z)d^2z) \cap \mathcal{O}(\mathbb{D}^*)_1$. Since $[\nu_i] \in W_{i-1} \cap W_i$, there exist $[\lambda_1], [\lambda_2] \in W_0$ such that $[\nu_i] = [\lambda_1 * \mu(t_{i-1})] = [\lambda_2 * \mu(t_i)]$. From the first equality it follows that $[\nu_i]$ has a representative $\nu_i \in L^2(\mathbb{D}^*, \rho(z)d^2z) \cap \mathcal{O}(\mathbb{D}^*)_1$, and the second equality implies that $[\mu(t_i)]$ has a representative $\mu_i \in L^2(\mathbb{D}^*, \rho(z)d^2z) \cap \mathcal{O}(\mathbb{D}^*)_1$. The assertion follows.
$\square$

LEMMA 3.4. *$T_0(1)$ is a subgroup of $T(1)$.*

PROOF. Every $[\lambda] \in T_0(1)$ has a representative $\lambda \in L^2(\mathbb{D}^*, \rho(z)d^2z) \cap \mathcal{O}(\mathbb{D}^*)_1$. Using Lemma 2.9 with $\mu = \lambda^{-1}$ we get by Corollary 3.2 that $[\lambda^{-1}] \in T_0(1)$. Now let $[\nu] \in T_0(1)$ with a representative $\nu \in L^2(\mathbb{D}^*, \rho(z)d^2z) \cap \mathcal{O}(\mathbb{D}^*)_1$. Using again Lemma 2.9 with $\mu = \nu$ we get that $[\lambda * \nu] \in T_0(1)$.
$\square$

The following lemmas and corollary are needed for the proof that $T_0(1)$ is a topological group. For every $[\mu] \in T(1)$ let
$$L_{[\mu]} : T(1) \to T(1), \quad [\nu] \mapsto [\mu * \nu],$$
be the left translation by $[\mu]$.

LEMMA 3.5. *Left translations are continuous on $T_0(1)$.*

PROOF. Let $[\mu], [\kappa] \in T_0(1)$ with representatives $\mu, \kappa \in L^2(\mathbb{D}^*, \rho(z)d^2z) \cap \mathcal{O}(\mathbb{D}^*)_1$, and let V_κ and $V_{\mu*\kappa}$ be coordinate charts introduced in Section 2.3. In terms of corresponding Bers coordinates the mapping $L_{[\mu]}$ takes the form

$$\beta \circ \Phi \circ R_{(\mu*\kappa)^{-1}} \circ L_\mu \circ R_\kappa \circ \Lambda : \beta(V_0) \subset A_2(\mathbb{D}) \to A_2(\mathbb{D}).$$

Since left translations commute with right translations, it simplifies to

$$\beta \circ \Phi \circ R_{\mu^{-1}} \circ L_\mu \circ \Lambda : \beta(V_0) \subset A_2(\mathbb{D}) \to A_2(\mathbb{D}),$$

which does not depend on $[\kappa]$. Thus to show that $L_{[\mu]}$ is continuous at $[\kappa] \in T_0(1)$ it is sufficient to show that the above mapping is continuous at $0 \in A_2(\mathbb{D})$. From Lemma 2.7 it follows that the mapping

$$R_{\mu^{-1}} : L^2(\mathbb{D}^*, \rho(z)d^2z) \cap L^\infty(\mathbb{D}^*)_1 \to L^2(\mathbb{D}^*, \rho(z)d^2z) \cap L^\infty(\mathbb{D}^*)_1$$

is continuous at μ, and from Lemma 2.9 it follows that the mapping

$$\beta \circ \Phi : L^2(\mathbb{D}^*, \rho(z)d^2z) \cap L^\infty(\mathbb{D}^*)_1 \to A_2(\mathbb{D})$$

is continuous at 0. What remains to show is that the mapping

$$L_\mu : H^{-1,1}(\mathbb{D}^*) \cap L^\infty(\mathbb{D}^*)_1 \to L^2(\mathbb{D}^*, \rho(z)d^2z),$$

$$\nu \mapsto \mu * \nu = \frac{(\mu \circ w_\nu)\overline{\frac{(w_\nu)_z}{(w_\nu)_z}} + \nu}{1 + (\mu \circ w_\nu)\bar{\nu}\overline{\frac{(w_\nu)_z}{(w_\nu)_z}}}$$

is continuous at 0.

For $\nu \in H^{-1,1}(\mathbb{D}^*) \cap L^\infty(\mathbb{D}^*)_1$ we have

$$L_\mu(\nu) - L_\mu(0) = \frac{(\mu \circ w_\nu)\overline{\frac{(w_\nu)_z}{(w_\nu)_z}} - \mu + \nu - (\mu \circ w_\nu)\mu\bar{\nu}\overline{\frac{(w_\nu)_z}{(w_\nu)_z}}}{1 + (\mu \circ w_\nu)\bar{\nu}\overline{\frac{(w_\nu)_z}{(w_\nu)_z}}}.$$

Since $\mu \in L^2(\mathbb{D}^*, \rho(z)d^2z)$, for every $\varepsilon > 0$ there exists $0 < r < 1$ such that

$$\iint_{\mathbb{D} \setminus \mathbb{D}_r} |\mu|^2 \rho(z)d^2z < \frac{\varepsilon^2}{16}.$$

It follows from Mori's theorem (see, e.g., [**Ahl87**]) that for every $0 < \delta_1 < 1$ there exists $0 < r_0 < 1$ such that for all $\|\nu\|_\infty < \delta_1$ and $r_0 \leq r < 1$,

$$w_\nu(\mathbb{D} \setminus \mathbb{D}_{r'}) \subset \mathbb{D} \setminus \mathbb{D}_r, \quad \text{for all} \quad r' \geq \frac{1+r}{2}.$$

Hence if $\|\nu\|_\infty < \min\{\delta, \delta_1\}$, where δ is as in Corollary 2.6, we have

$$\iint_\mathbb{D} \left| \mu \circ w_\nu \overline{\frac{(w_\nu)_z}{(w_\nu)_z}} - \mu \right|^2 \rho(z)d^2z \leq 2\iint_{\mathbb{D} \setminus \mathbb{D}_{r'}} \left| \mu \circ w_\nu \overline{\frac{(w_\nu)_z}{(w_\nu)_z}} \right|^2 \rho(z)d^2z$$

$$+ 2\iint_{\mathbb{D} \setminus \mathbb{D}_{r'}} |\mu|^2 \rho(z)d^2z + \iint_{\mathbb{D}_{r'}} \left| \mu \circ w_\nu \overline{\frac{(w_\nu)_z}{(w_\nu)_z}} - \mu \right|^2 \rho(z)d^2z$$

$$\leq \frac{3\varepsilon^2}{8} + \iint_{\mathbb{D}_{r'}} \left| \mu \circ w_\nu \overline{\frac{(w_\nu)_z}{(w_\nu)_z}} - \mu \right|^2 \rho(z)d^2z.$$

Since $w_\nu(z) \to z$, $(w_\nu)_z(z) \to 1$ as $\|\nu\|_\infty \to 0$ uniformly for all $z \in \mathbb{D}_{r'}$, we can choose δ_2 such that for all $\|\nu\|_\infty < \delta_2$, the second integral is less than $\frac{\varepsilon^2}{8}$. Finally, let $\delta_3 \leq \sqrt{\frac{4\pi}{3}} \min\{\delta, \delta_1, \delta_2\}$, then for all $\|\nu\|_2 < \delta_3$, we have

$$\left\| L_\mu(\nu) - L_\mu(0) \right\|_2 < \frac{1}{1 - \sqrt{\frac{3}{4\pi}}\delta_3} \left(\frac{\varepsilon}{\sqrt{2}} + 2\delta_3 \right),$$

which is less than ε for δ_3 sufficiently small. $\square$

Notice that what we actually prove is the following

COROLLARY 3.6. *For every $[\mu] \in T_0(1)$ the adjoint mapping*

$$T_0(1) \ni [\nu] \mapsto [\mu * \nu * \mu^{-1}] \in T_0(1)$$

is continuous at the origin.

LEMMA 3.7. *The following statements hold.*
  (i) *The inverse mapping $T_0(1) \ni [\mu] \mapsto [\mu^{-1}] \in T_0(1)$ is continuous at the origin.*
  (ii) *The group multiplication $T_0(1) \times T_0(1) \ni ([\mu], [\nu]) \mapsto [\mu * \nu] \in T_0(1)$ is continuous at the origin.*

PROOF. In coordinate chart V_0 the inverse mapping takes the form

$$A_2(\mathbb{D}) \supset \beta(V_0) \ni \phi \mapsto \beta \circ \Phi((\Lambda\phi)^{-1}) \in A_2(\mathbb{D}).$$

By Lemma 2.9, it is continuous at 0. This proves (i). In coordinate chart V_0 the group multiplication takes the form

$$\beta(V_0) \times \beta(V_0) \ni (\phi_1, \phi_2) \mapsto \beta \circ \Phi(\Lambda(\phi_1) * \Lambda(\phi_2)) \in A_2(\mathbb{D}).$$

Its continuity at $(0,0) \in A_2(\mathbb{D}) \times A_2(\mathbb{D})$ follows from Lemma 2.9 and the proof of Lemma 3.5. $\square$

THEOREM 3.8. *The Hilbert manifold $T_0(1)$ is a topological group.*

PROOF. We need to prove that the map

$$T_0(1) \times T_0(1) \ni ([\mu], [\nu]) \mapsto [\mu * \nu^{-1}] \in T_0(1)$$

is continuous at every point $([\mu], [\nu]) \in T_0(1) \times T_0(1)$ with representatives $\mu, \nu \in L^2(\mathbb{D}^*, \rho(z)d^2z) \cap \mathcal{O}(\mathbb{D}^*)_1$. In coordinate charts V_μ, V_ν and V_κ, where $\kappa = \mu * \nu^{-1}$, this map takes the form

$$\beta(V_0) \times \beta(V_0) \ni (\phi_1, \phi_2) \mapsto \beta \circ \Phi(\Lambda(\phi_1) * \kappa * \Lambda(\phi_2)^{-1} * \kappa^{-1}) \in A_2(\mathbb{D}).$$

Its continuity at $(0,0) \in A_2(\mathbb{D}) \times A_2(\mathbb{D})$ follows from Corollary 3.6, Lemma 3.7 and Lemma 2.9. $\square$

Finally, we show that $\mathcal{T}_0(1)$ is also a topological group.

PROPOSITION 3.9. *The Hilbert manifold $\mathcal{T}_0(1)$ is a topological group.*

PROOF. We have to prove that the multiplication map
$$\mathcal{T}_0(1) \times \mathcal{T}_0(1) \ni (\gamma_1, \gamma_2) \mapsto \gamma_1 * \gamma_2^{-1} \in \mathcal{T}_0(1)$$
is continuous. Setting $\gamma_1 = ([\nu], \zeta)$ and $\gamma_2 = ([\mu], w)$, we get from (1.13) that $\gamma_1 * \gamma_2^{-1} = ([\lambda], z)$, where
$$\lambda = (\sigma_u^{-1})^*(\nu * \mu^{-1}), \quad z = (w^\lambda \circ \gamma_1 \circ (w^\nu)^{-1})(\zeta),$$
and
$$\gamma_2 = \sigma_u \circ w_\mu = g_\mu^{-1} \circ f^\mu, \quad u = g_\mu^{-1}(w).$$
It easily follows from the proof of Lemma 3.5 that the mapping
$$\text{Möb}(S^1) \times T(1) \ni (\sigma, [\mu]) \mapsto [(\sigma^{-1})^*(\mu)] \in T(1)$$
is continuous. Now the map $\mathcal{T}(1) \ni ([\mu], w) \mapsto g_\mu^{-1}(w) \in \mathbb{D}^*$ depends continuously on $[\mu]$ (uniformly when w varies on compact subsets), so using that $T_0(1)$ is a topological group, we conclude that $[\lambda] \in T_0(1)$ depends continuously on $\gamma_1, \gamma_2 \in \mathcal{T}_0(1)$. Finally, by the result of Bers [**Ber73**] (which we already have used in the proof of Lemma 5.4) we get that $z \in \mathbb{D}^*$ also depends continuously on γ_1 and γ_2. □

4. Velling-Kirillov and Weil-Petersson metrics

4.1. Velling-Kirillov metric on the universal Teichmüller curve. The Velling-Kirillov metric is a right-invariant Hermitian metric on $\mathcal{T}(1)$, defined at the origin of $\mathcal{T}(1)$ by

(4.1) $$\|\mathbf{v}\|_{VK}^2 = \sum_{n=1}^\infty n |c_n|^2,$$

where

$$\mathbf{v} = \frac{\mathbf{u} - iJ\mathbf{u}}{2} \quad \text{and} \quad \mathbf{u} = \sum_{n \in \mathbb{Z} \setminus \{0\}} c_n e^{in\theta} \frac{d}{d\theta} \in T_0^{\mathbb{R}} S^1 \backslash \text{Homeo}_{qs}(S^1).$$

The convergence of the series is guaranteed by the property **TS4** (with $s = 1/2$) in Section 1.2.3. The Velling-Kirillov metric is a smooth right-invariant Kähler metric on the complex Banach manifold $\mathcal{T}(1)$. Its symplectic form ω_{VK} at the origin of $\mathcal{T}(1)$ is given by

$$\omega_{VK}(\mathbf{v}, \bar{\mathbf{v}}) = \frac{i}{2} \|\mathbf{v}\|_{VK}^2,$$

In the following section we prove that the Velling-Kirillov metric is real-analytic on $\mathcal{T}(1)$ by presenting its real-analytic Kähler potential.

REMARK 4.1. For the homogeneous space $S^1 \backslash \text{Diff}_+(S^1)$ the metric was introduced in this form by A.A. Kirillov [**Kir87**] and it has been studied by A.A. Kirillov and D. Yuriev. In particular, in [**KY87**] it was shown to be Kähler. In [**Vel**], J. Velling has introduced a Hermitian metric for the space $\mathcal{T}(1)$ using arguments from geometric function theory. In [**Teo04**], Kirillov's definition was extended to $\mathcal{T}(1)$ and it was shown that the resulting metric coincides with the metric introduced by Velling. The Velling-Kirillov metric is the unique right-invariant Kähler metric on the universal Teichmüller curve $\mathcal{T}(1)$ [**Teo04**].

4.2. Weil-Petersson metric on the universal Teichmüller space. In this section we consider $T(1)$ as a Hilbert manifold. The Weil-Petersson metric on $T(1)$ is a Hermitian metric defined by the Hilbert space inner product on tangent spaces, which are identified with the Hilbert space $H^{-1,1}(\mathbb{D}^*)$ by right translations (see Section 2.3). Thus the Weil-Petersson metric is a right-invariant metric on $T(1)$ defined at the origin of $T(1)$ by

$$(4.2) \qquad \langle \mu, \nu \rangle_{WP} = \iint_{\mathbb{D}^*} \mu \bar{\nu} \rho(z) d^2 z, \quad \mu, \nu \in H^{-1,1}(\mathbb{D}^*) = T_0 T(1).$$

To every $\mu \in H^{-1,1}(\mathbb{D}^*)$ there corresponds a vector field $\frac{\partial}{\partial \varepsilon_\mu}$ over V_0, given by (1.5)-(1.7). We set for every $\kappa \in V_0$,

$$(4.3) \qquad g_{\mu\bar{\nu}}(\kappa) = \left\langle \frac{\partial}{\partial \varepsilon_\mu}\Big|_\kappa, \frac{\partial}{\partial \varepsilon_\nu}\Big|_\kappa \right\rangle_{WP} = \iint_{\mathbb{D}^*} P(R(\mu,\kappa)) \overline{P(R(\nu,\kappa))} \rho(z) d^2 z.$$

This formula explicitly defines the Weil-Petersson metric on the coordinate chart V_0. The Weil-Petersson metric extends to other charts V_μ by right translations.

The following statement is an easy consequence of Lemma 2.5.

LEMMA 4.2. *The Weil-Petersson metric is continuous on $T(1)$.*

PROOF. As it follows from Corollary 2.12, it is sufficient to prove that for every $\mu \in H^{-1,1}(\mathbb{D}^*)$ the function $g_{\mu\bar{\mu}}$ is continuous on V_0 at 0. Since the embedding $V_0 \hookrightarrow U_0$ is continuous by Lemma 2.1, it is sufficient to prove that the function $g_{\mu\bar{\mu}}$ is defined on a neighborhood of 0 in U_0 and is continuous at 0.

Since the projector P is norm-decreasing,

$$g_{\mu\bar{\mu}}(\kappa) = \iint_{\mathbb{D}^*} P(R(\mu,\kappa)) \overline{P(R(\mu,\kappa))} \rho(z) d^2 z$$

$$\leq \iint_{\mathbb{D}^*} R(\mu,\kappa) \overline{R(\mu,\kappa)} \rho(z) d^2 z$$

$$= 4 \iint_{\mathbb{D}^*} \frac{|\mu|^2}{1-|\kappa|^2} \frac{|(w_\kappa)_z|^2}{(1-|w_\kappa|^2)^2} d^2 z.$$

According to Lemma 2.5, for every $\varepsilon > 0$ there exists $0 < \delta < 1$ such that for all $\kappa \in U_0$ satisfying $\|\kappa\|_\infty < \delta$ we have

$$\left| g_{\mu\bar{\mu}}(\kappa) - g_{\mu\bar{\mu}}(0) \right| \leq 4 \iint_{\mathbb{D}^*} \frac{|\mu|^2}{1-|\kappa|^2} \left| \frac{|(w_\kappa)_z|^2}{(1-|w_\kappa|^2)^2} - \frac{1}{(1-|z|^2)^2} \right|$$

$$+ \frac{|\mu|^2}{(1-|z|^2)^2} \left(\frac{1}{1-|\kappa|^2} - 1 \right) d^2 z$$

$$\leq \frac{\varepsilon + \delta^2}{1 - \delta^2} \iint_{\mathbb{D}} |\mu|^2 \rho d^2 z.$$

Thus, for δ small enough $\left| g_{\mu\bar{\mu}}(\kappa) - g_{\mu\bar{\mu}}(0) \right| \leq 2\varepsilon \|\mu\|_2^2$. □

REMARK 4.3. Using the basic properties of the q.c. mappings, it can be shown that the Weil-Petersson metric is real-analytic on $T(1)$. In fact, it is sufficient to prove that for every $\mu, \nu \in H^{-1,1}(\mathbb{D}^*)$ the mapping $V_0 \ni \kappa \mapsto g_{\mu\bar{\nu}}(\kappa) \in \mathbb{C}$ is real-analytic on V_0. Since this result will not be used later, we omit the proof. Explicit curvature computations in Section 7 will show that the Weil-Petersson metric on $T(1)$ is twice differentiable.

We will prove in Section 7 that the Weil-Petersson metric is Kähler. Its symplectic form ω_{WP} is a right-invariant $(1,1)$ form on the Hilbert manifold $T(1)$. At the origin of $T(1)$,
$$\omega_{WP}(\mu, \bar{\nu}) = \tfrac{i}{2} \langle \mu, \nu \rangle_{WP}, \quad \mu, \nu \in H^{-1,1}(\mathbb{D}^*).$$

REMARK 4.4. The Weil-Petersson metric on the distribution $\mathfrak{D}_T$ (without defining the Hilbert manifold structure) was introduced by S. Nag and A. Verjovsky [**NV90**] as a direct generalization of the Weil-Petersson metric on the finite-dimensional Teichmüller spaces. It was proved in [**NV90**] that the embedding $\text{Möb}(S^1)\backslash \text{Diff}_+(S^1) \hookrightarrow T(1)$ is holomorphic and the pull-back of the Weil-Petersson metric on the distribution $\mathfrak{D}_T$ coincides, up to a constant, with the right invariant Kähler metric introduced by Kirillov [**Kir87**] by the orbit method. At the tangent space of the origin the latter metric is defined by (cf. (4.1))
$$\|\mathbf{v}\|^2 = \sum_{n=2}^{\infty} (n^3 - n)|c_n|^2,$$
where
$$\mathbf{v} = \frac{\mathbf{u} - iJ\mathbf{u}}{2} \quad \text{and} \quad \mathbf{u} = \sum_{n \in \mathbb{Z}\setminus\{0, \pm 1\}} c_n e^{in\theta} \frac{d}{d\theta} \in T_0^{\mathbb{R}} \text{Möb}(S^1)\backslash \text{Diff}_+(S^1).$$

5. Characteristic forms of the universal Teichmüller curve

Let $V = T_v \mathcal{T}(1)$ be the vertical tangent bundle of the fibration
$$\pi : \mathcal{T}(1) \to T(1).$$
It is a holomorphic line bundle over the complex Banach manifold $\mathcal{T}(1)$, the fiber over a point $([\mu], z) \in \mathcal{T}(1)$ is the holomorphic tangent bundle to the quasi-disk $w^\mu(\mathbb{D}^*)$. The hyperbolic metric on $w^\mu(\mathbb{D}^*)$ defines a Hermitian metric on V, and we denote by $c_1(V)$ the first Chern form of V corresponding to this metric.

The Hilbert manifold structure on $T(1)$ naturally induces a Hilbert manifold structure on $\mathcal{T}(1)$, such that the projection $\pi : \mathcal{T}(1) \to T(1)$ is a holomorphic mapping of Hilbert manifolds. Similar to the Hilbert manifold $T(1)$, the Hilbert manifold $\mathcal{T}(1)$ is also a disjoint union of uncountably many components. We will prove in Appendix A that the component containing the identity $\mathcal{T}_0(1)$ is a topological group.

The vertical tangent bundle is also a holomorphic line bundle over the Hilbert manifold $\mathcal{T}(1)$. We will continue to denote corresponding line bundle by V, and by $c_1(V)$ — the first Chern form corresponding to the hyperbolic metric on the fibers, specifying explicitly which topology we are using. Since the Hilbert manifold topology is stronger than the Banach manifold topology, the form $c_1(V)$ for the Banach manifold structure on $\mathcal{T}(1)$ naturally restricts onto $\mathcal{T}(1)$ considered as a Hilbert manifold.

Similar to Wolpert's work [**Wol86**] on finite dimensional Teichmüller spaces, we define the analogs of Mumford-Morita-Miller characteristic forms as the following

5. CHARACTERISTIC FORMS OF THE UNIVERSAL TEICHMÜLLER CURVE

(n, n)-forms on the Hilbert manifold $T(1)$,

$$\kappa_n = (-1)^{n+1}\pi_*(c_1(V)^{n+1}), \tag{5.1}$$

where $\pi_* : \Omega^*(\mathcal{T}(1)) \to \Omega^{*-2}(T(1))$ is the operation of "integration over the fibers" of the fibration $\pi : \mathcal{T}(1) \to T(1)$. As we will see in Section 5.3, it is the passage from the Banach manifold structure to the Hilbert manifold structure that makes the operation π_* well-defined (i.e., the integrals over non-compact fibers become convergent).

5.1. The form $c_1(V)$ as Velling-Kirillov symplectic form. In this section we work with the Banach manifold structure on $\mathcal{T}(1)$. Let z be the complex coordinate on $\hat{\mathbb{C}} \setminus \{0\}$. The assignment $\mathcal{T}(1) \ni ([\mu], z) \mapsto -z^2 \partial_z$ defines a holomorphic section of the line bundle V over $\mathcal{T}(1)$ [4]. The hyperbolic metric on $w^\mu(\mathbb{D}^*)$ is the pull-back of the hyperbolic metric on $\mathbb{D}^*$ by the conformal map g_μ^{-1}, so that

$$\left\| z^2 \partial_z \right\|^2_{([\mu],z)} = \frac{4|z^2(g_\mu^{-1})'(z)|^2}{(|g_\mu^{-1}(z)|^2 - 1)^2}.$$

The first Chern form of the line bundle V is

$$c_1(V) = \frac{i}{2\pi} \Theta = \frac{i}{2\pi} \bar{\partial}\partial \log \left\| z^2 \partial_z \right\|^2,$$

where ∂ and $\bar{\partial}$ are, respectively, the $(1,0)$ and $(0,1)$ components of the de Rham differential on $\mathcal{T}(1)$.

Let

$$K = \log \left\| z^2 \partial_z \right\|_{([\mu],z)} - \log 2.$$

LEMMA 5.1. *The function $K : \mathcal{T}(1) \to \mathbb{R}$ is real-analytic. Under the correspondence $\mathcal{T}(1) \ni ([\mu], z) \mapsto \mathrm{g} \in S^1 \backslash \mathrm{Homeo}_{qs}(S^1)$, where $\gamma = (g^{-1} \circ f)|_{S^1}$,*

$$K(\gamma) = \log |g'(\infty)|.$$

PROOF. Using the formulas $g = \lambda_w \circ g_\mu \circ \sigma_w^{-1}$ and $w = g_\mu^{-1}(z)$ from Section 1.2.1, it is straightforward to compute

$$g'(\infty) = \frac{z^2(g_\mu^{-1})'(z)(1-\bar{w})}{(|w|^2-1)(1-w)}.$$

Now it easily follows from the general properties of q.c. mappings that for $z \in \mathbb{D}^*$ the functional $T(1) \ni [\mu] \mapsto g_\mu^{-1}(z) \in \mathbb{C}$ is real-analytic so that $|g'(\infty)|$ is a real-analytic function on $\mathcal{T}(1)$. □

REMARK 5.2. The quantity $|g'(\infty)|$ is the capacity of the quasi-circle $g(S^1)$ corresponding to $\gamma \in \mathrm{Homeo}_{qs}(S^1)$.

THEOREM 5.3. *The first Chern form of the vertical tangent bundle to the universal Teichmüller curve $\mathcal{T}(1)$ is proportional to the symplectic form of the Velling-Kirillov metric,*

$$c_1(V) = -\frac{2}{\pi} \omega_{VK}.$$

Equivalently, the function K is a Kähler potential for the Velling-Kirillov metric on $\mathcal{T}(1)$.

PROOF. It is based on the following lemmas.

---

[4] Under the conformal map $z \mapsto \frac{1}{z}$ the vector field $-z^2 \partial_z \mapsto \partial_z$.

LEMMA 5.4. *The $(1,1)$-form $c_1(V)$ on $\mathcal{T}(1)$ is right-invariant.*

PROOF. We need to prove that for every $g_0 \in S^1\backslash\mathrm{Homeo}(S^1) \cong \mathcal{T}(1)$ the difference $R_{\gamma_0}^* K - K$, where $R_{\gamma_0} : \mathcal{T}(1) \to \mathcal{T}(1)$ is the right translation by γ_0 and $(R_{\gamma_0}^* K)(\gamma) = K(\gamma \circ \gamma_0)$, is a harmonic function on $\mathcal{T}(1)$.

For every $\gamma = g^{-1} \circ f \in \mathcal{T}(1)$ let $\tilde{\gamma} = \gamma \circ \gamma_0 = \tilde{g}^{-1} \circ \tilde{f}$. Since $\tilde{g} = \tilde{f} \circ g_0^{-1} \circ f^{-1} \circ g$, we have

$$\left(R_{\gamma_0}^* K - K\right)(\gamma) = \log|\tilde{g}'(\infty)| - \log|g'(\infty)| = \log|(\tilde{f} \circ \gamma_0^{-1} \circ f^{-1})'(\infty)|.$$

In [**Ber73**], Bers has proved that the function

$$(\gamma, z) \mapsto \left(\tilde{f} \circ \gamma_0^{-1} \circ f^{-1}\right)(z) = h(\gamma, z)$$

depends holomorphically on γ and z, which implies that $\left(\tilde{f} \circ \gamma_0^{-1} \circ f^{-1}\right)'(\infty)$ depends holomorphically on γ and our assertion follows. □

LEMMA 5.5. *Let $\gamma = g^{-1} \circ f \in \mathcal{T}(1)$, where $f|_{\mathbb{D}}(z) = \sum_{n=0}^{\infty} a_n z^{n+1}$, $a_0 = 1$, and $g|_{\mathbb{D}^*}(z) = \sum_{n=0}^{\infty} b_n z^{1-n}$. Then*

$$(5.2) \qquad |b_0|^2 = \sum_{n=0}^{\infty}(n+1)|a_n|^2 + \sum_{n=1}^{\infty}(n-1)|b_n|^2.$$

PROOF. Evaluate the Euclidean area of the domain $\Omega = f(\mathbb{D}) = g(\mathbb{D})$ in two different ways. First,

$$A_E(\Omega) = \lim_{r \to 1^-} \iint_{f(\mathbb{D}_r)} d^2 z = \lim_{r \to 1^-} \iint_{\mathbb{D}_r} |f'|^2 d^2 z = \pi \sum_{n=0}^{\infty}(n+1)|a_n|^2,$$

where $\mathbb{D}_r = \{z \in \mathbb{C} : |z| < r\}$. On the other hand, the classical area theorem gives

$$A_E(\Omega) = \pi \sum_{n=0}^{\infty}(1-n)|b_n|^2,$$

and we obtain (5.2). □

Now we complete the proof of the theorem. Let

$$\mathrm{u} = \sum_{n \in \mathbb{Z}\backslash\{0\}} c_n e^{in\theta} \frac{d}{d\theta} \in T_0^{\mathbb{R}} \mathcal{T}(1),$$

and let $\gamma_t = g_t^{-1} \circ f^t$, $\gamma_0 = \mathrm{id}$, be the corresponding smooth curve in $\mathcal{T}(1)$. Using notations from the previous lemma, differentiate the relation (5.2) with respect to t and set $t = 0$, and using $b_0(0) = 1$, $a_n(0) = b_n(0) = 0$ for $n \geq 1$, we get

$$(5.3) \qquad \dot{b}_0 + \bar{\dot{b}}_0 = 0.$$

Here we denote

$$\dot{a}_n(\mathrm{u}) = \dot{a}_n = \frac{da_n}{dt}(0) \quad \text{and} \quad \dot{b}_n(\mathrm{u}) = \dot{b}_n = \frac{db_n}{dt}(0).$$

Differentiating (5.2) twice with respect to t and setting $t=0$ we get

$$\frac{d^2 b_0}{dt^2}(0) + \frac{d^2 \bar{b}_0}{dt^2}(0) + 2\left|\frac{db_0}{dt}(0)\right|^2 = 2\sum_{n=1}^\infty (n+1)|\dot{a}_n|^2 + 2\sum_{n=1}^\infty (n-1)|\dot{b}_n|^2$$
$$= 4\sum_{n=1}^\infty n|\dot{a}_n|^2,$$

where we have also used the property **TS1** in Section 1.2.3. Since $g'_t(\infty) = b_0(t)$, using (5.3) we get

$$\frac{d^2}{dt^2}\log|g'_t(\infty)|\big|_{t=0} = 2\sum_{n=1}^\infty n|\dot{a}_n|^2.$$

Let $\mathrm{v} = \frac{1}{2}(\mathrm{u} - iJ\mathrm{u})$ be the holomorphic tangent vector to $\mathcal{T}(1)$. Since $\dot{a}_n(J\mathrm{u}) = i\dot{a}_n(\mathrm{u})$, using (1.16) we finally get

$$(5.4) \qquad (\partial\bar{\partial}K)(\mathrm{v},\bar{\mathrm{v}}) = \frac{1}{2}\sum_{n=1}^\infty n\left(|\dot{a}_n(\mathrm{u})|^2 + |\dot{a}_n(J\mathrm{u})|^2\right) = \sum_{n=1}^\infty n|\dot{a}_n(\mathrm{u})|^2.$$

This proves that $\Theta = 4i\omega_{VK}$ at the origin of $\mathcal{T}(1)$. Since both these $(1,1)$-forms on $\mathcal{T}(1)$ are right-invariant, the assertion follows. □

REMARK 5.6. In [**KY87**], A.A. Kirillov and D. Yuriev have stated that the function K, restricted to the space $S^1\backslash\mathrm{Diff}_+(S^1)$, is a Kähler potential of the Velling-Kirillov metric. Theorem 5.3 extends this result to $\mathcal{T}(1) \simeq S^1\backslash\mathrm{Homeo}_{qs}(S^1)$ and gives its geometric interpretation.

5.2. The Chern form $c_1(V)$ and the resolvent kernel. Let $\mu \in \Omega^{-1,1}(\mathbb{D}^*)$ be a horizontal holomorphic tangent vector to $\mathcal{T}(1)$ at the origin. According to the property **TV1** in Section 1.2.2, the vector field τ_μ — the horizontal lift of the vector field $\frac{\partial}{\partial\varepsilon_\mu}$ on $U_0 \subset T(1)$ to the point $(0,z) \in \pi^{-1}(0)$, is identified with $(\sigma_z^{-1})^*\mu \in \Omega^{-1,1}(\mathbb{D}^*)$.

PROPOSITION 5.7.   (i) *On the fiber $\pi^{-1}(0) \subset \mathcal{T}(1)$ the Velling-Kirillov metric is given by*

$$\langle\partial_z, \partial_z\rangle_{VK}(0,z) = \frac{1}{(1-|z|^2)^2},$$
$$\langle\partial_z, \tau_\mu\rangle_{VK}(0,z) = 0,$$
$$\langle\tau_\mu, \tau_\mu\rangle_{VK}(0,z) = \frac{1}{2}\iint_{\mathbb{D}^*} G(z,u)|\mu(u)|^2\rho(u)d^2u.$$

(ii) *On the fiber $\pi^{-1}(0) \subset \mathcal{T}(1)$ the $(1,1)$-form Θ is given by*

$$\Theta(\partial_z, \partial_{\bar{z}})(0,z) = -\frac{2}{(1-|z|^2)^2},$$
$$\Theta(\partial_z, \tau_{\bar{\mu}})(0,z) = 0,$$
$$\Theta(\tau_\mu, \tau_{\bar{\mu}})(0,z) = -\iint_{\mathbb{D}^*} G(z,u)|\mu(u)|^2\rho(u)d^2u.$$

(iii) *The vertical holomorphic tangent bundle $V \to \mathcal{T}(1)$ of the fibration $\pi : \mathcal{T}(1) \to T(1)$ is a negative line bundle.*

PROOF. It follows from the property **TV2** in Section 1.2.2 that the vector field ∂_z at $(0, z) \in \pi^{-1}(0)$ corresponds to the tangent vector

$$u = \sum_{n \in \mathbb{Z} \setminus \{0\}} c_n e^{in\theta} \frac{d}{d\theta} \in T_0^{\mathbb{R}} S^1 \backslash \mathrm{Homeo}_{qs}(S^1)$$

with $c_1 = \frac{1-\bar{z}}{(1-z)(1-|z|^2)}$ and $c_n = 0$ for $n \geq 2$. This proves the first formula in part (i). The second formula follows from the fact that, according to the property **TS1** in Section 1.2.3, the tangent vector $u \in T_0^{\mathbb{R}} S^1 \backslash \mathrm{Homeo}_{qs}(S^1)$ which corresponds to the horizontal lift τ_μ of the vector field $\frac{\partial}{\partial \varepsilon_\mu}$ to $(0, z) \in \pi^{-1}(0)$, has $c_1 = 0$. The last formula follows from the following lemma.

LEMMA 5.8. *Let $\mu \in \Omega^{-1,1}(\mathbb{D}^*)$ and*

$$\phi(z) = D_0\beta(\mu)(z) = \sum_{n=2}^\infty (n^3 - n)a_n z^{n-2} \in A_\infty(\mathbb{D}).$$

Then

$$\iint_{\mathbb{D}^*} G(z, u)|\mu(u)|^2 \rho(u) d^2 u = 2 \sum_{n=2}^\infty n|a_n^z|^2,$$

where

$$(\sigma_z^{-1})^*(\phi)(u) = \sum_{n=2}^\infty (n^3 - n)a_n^z u^{n-2}$$

is the power series expansion of $(\sigma_z^{-1})^*(\phi) = \phi \circ \sigma_z^{-1}\left((\sigma_z^{-1})'\right)^2 \in A_\infty(\mathbb{D})$.

PROOF. Since $G(z, u)$ is a point-pair invariant, it is sufficient to prove the formula for $z = \infty$. In this case, using the relation $G(\infty, u) = G(0, 1/\bar{u})$ between the resolvent kernels on $\mathbb{D}^*$ and $\mathbb{D}$ and the formula $\mu = \Lambda(\phi)$ we get

$$\iint_{\mathbb{D}^*} G(\infty, u)|\mu(u)|^2 \rho(u) d^2 u = \iint_{\mathbb{D}} G(0, u)(1 - |u|^2)^2 |\phi(u)|^2 d^2 u.$$

It follows from the explicit formula (1.19) that

$$(1 - |u|^2)^2 G(0, u) = h(r) = \left(\frac{1}{2\pi} \frac{1+r^2}{1-r^2} \log \frac{1}{r^2} - \frac{1}{\pi}\right)(1 - r^2)^2, \quad r = |u|.$$

Since h is an integrable function on $[0, 1)$, we have

$$\iint_{\mathbb{D}} G(0, u)(1 - |u|^2)^2 |\phi(u)|^2 d^2 u = 2\pi \sum_{n=2}^\infty (n^3 - n)^2 |a_n|^2 \int_0^1 h(r) r^{2n-4} r dr.$$

A straightforward computation gives

$$\int_0^1 h(r) r^{2n-4} r dr = \frac{1}{\pi} \frac{1}{(n^3 - n)(n^2 - 1)},$$

which proves the lemma. □

Using this Lemma, the property **TV1** in Section 1.2.2 and the property **TS2** in Section 1.2.3, we get the third formula in part (i). Now part (ii) follows from Theorem 5.3, and part (iii) follows from part (ii) and the property **RK2** in Section 1.4. □

REMARK 5.9. Parts (ii) and (iii) of Proposition 5.7 generalize Wolpert's computation of the $(1,1)$-form Θ for finite-dimensional Teichmüller spaces (see Theorem 5.5 and formula (5.3) in [**Wol86**]).

5.3. Mumford-Morita-Miller characteristic forms. In this section we consider $\pi : \mathcal{T}(1) \to T(1)$ as a holomorphic fibration of Hilbert manifolds and evaluate the Mumford-Morita-Miller forms κ_n on $T(1)$.

THEOREM 5.10. *The characteristic forms κ_n are right-invariant on the Hilbert manifold $T(1)$ and for $\mu_1, \ldots, \mu_n, \nu_1, \ldots, \nu_n \in H^{-1,1}(\mathbb{D}^*) \simeq T_0 T(1)$*

$$\kappa_n(\mu_1, \ldots, \mu_n, \bar{\nu}_1, \ldots, \bar{\nu}_n)$$
$$= \frac{i^n (n+1)!}{(2\pi)^{n+1}} \sum_{\sigma \in S_n} sgn(\sigma) \iint_{\mathbb{D}^*} G\left(\mu_1 \bar{\nu}_{\sigma(1)}\right) \ldots G\left(\mu_n \bar{\nu}_{\sigma(n)}\right) \rho(z) d^2 z,$$

where the sum goes over the permutation group S_n on n elements and $sgn(\sigma)$ is the sign of the permutation σ.

PROOF. It is straightforward computation of the integral

$$\kappa_n(\mu_1, \ldots, \mu_n, \bar{\nu}_1, \ldots, \bar{\nu}_n)$$
$$= \left(\frac{-i}{2\pi}\right)^{n+1} \iint_{\mathbb{D}^*} \Theta^{n+1}\left(\partial_z, \partial_{\bar{z}}, \tau_{\mu_1}, \tau_{\bar{\nu}_1}, \ldots, \tau_{\mu_n}, \tau_{\bar{\nu}_n}\right) dz \wedge d\bar{z}$$

using Part (ii) of Proposition 5.7. We need only to verify that the integral is convergent. This follows from the properties of the resolvent kernel in Section 1.4. Indeed, the property **RK3** assures that $G(\mu\bar{\nu})$ is bounded on $\mathbb{D}^*$ for $\mu, \nu \in \Omega^{-1,1}(\mathbb{D}^*)$, and properties **RK2** and **RK4** imply that for $\mu, \nu \in H^{-1,1}(\mathbb{D}^*)$,

$$\left|\iint_{\mathbb{D}^*} G(\mu\bar{\nu})\rho(z)d^2z\right| \leq \iint_{\mathbb{D}^*} \iint_{\mathbb{D}^*} G(z,u)|\mu(u)\nu(u)|\rho(u)\rho(z)d^2u d^2z$$
$$= \iint_{\mathbb{D}^*} \iint_{\mathbb{D}^*} G(z,u)|\mu(u)\nu(u)|\rho(z)\rho(u)d^2z d^2u$$
$$= \iint_{\mathbb{D}^*} |\mu(u)\nu(u)|\rho(u)d^2u < \infty.$$

□

COROLLARY 5.11. *On the Hilbert manifold $T(1)$*

$$\kappa_1 = \tfrac{1}{\pi^2}\, \omega_{WP}.$$

PROOF. We have, using again the property **RK4** in Section 1.4,

$$\kappa_1(\mu, \bar{\nu}) = \frac{i}{2\pi^2} \iint_{\mathbb{D}^*} G(\mu\bar{\nu})\rho(z)d^2z = \frac{i}{2\pi^2} \iint_{\mathbb{D}^*} \mu(z)\overline{\nu(z)}\rho(z)d^2z$$

$$= \frac{1}{\pi^2} \omega_{WP}(\mu, \bar{\nu}).$$

□

REMARK 5.12. Combining Corollary 5.11, Part (ii) of Proposition 5.7 and Theorem 5.3, we get another proof of Theorem 4.3 in [**Teo04**].

REMARK 5.13. Theorem 5.10 generalizes Wolpert's result for finite-dimensional Teichmüller spaces (see Lemma 5.9 and Lemma 5.10 in [**Wol86**]) to the universal Teichmüller space.

6. First and second variations of the hyperbolic metric

Here we present a concise formula for the second variation of the hyperbolic metric in terms of the resolvent kernel. We are using the model $\mathbb{H}^2 \simeq \mathbb{U}$, so that the density of the hyperbolic metric is the $(1,1)$ – tensor $\rho(z) = y^{-2}$ on $\mathbb{U}$.

6.1. The first variation. It is a classical result of Ahlfors [**Ahl61**] that the first variation of the hyperbolic metric at the origin of $T(1)$ is identically zero.

LEMMA 6.1. *For every* $\mu \in \Omega^{-1,1}(\mathbb{U})$,

$$L_\mu \rho = 0.$$

PROOF. Since

$$\rho^{\varepsilon\mu} = w_{\varepsilon\mu}^*(\rho) = -\frac{4|(w_{\varepsilon\mu})_z(z)|^2}{(w_{\varepsilon\mu}(z) - \overline{w_{\varepsilon\mu}(z)})^2},$$

we have

$$L_\mu \rho(z) = \frac{\partial}{\partial \varepsilon}\bigg|_{\varepsilon=0} \rho^{\varepsilon\mu}(z) = -4\frac{F_z(z) + \overline{\Phi'(z)}}{(z-\bar{z})^2} + 8\frac{F(z) - \overline{\Phi(z)}}{(z-\bar{z})^3},$$

where $F = F[\mu], \Phi = \Phi[\mu]$, and the result follows from Lemma 1.13. □

REMARK 6.2. For the case $\mu \in \Omega^{-1,1}(\mathbb{U}, \Gamma)$, where Γ is a cofinite Fuchsian group, another proof of the Ahlfors result was given by Wolpert [**Wol86**].

6.2. The second variation. Set

$$L_\mu L_{\bar{\mu}} \rho = \frac{\partial^2}{\partial\varepsilon\partial\bar{\varepsilon}}\bigg|_{\varepsilon=0} \rho^{\varepsilon\mu}.$$

We have

PROPOSITION 6.3. *For every* $\mu \in \Omega^{-1,1}(\mathbb{U})$,

$$L_\mu L_{\bar{\mu}} \rho = \rho G(|\mu|^2).$$

6. FIRST AND SECOND VARIATIONS OF THE HYPERBOLIC METRIC

PROOF. Using the representation

$$\rho^{\varepsilon\mu}(z) = -4K_{\varepsilon\mu}(z, \bar{z}) \tag{6.1}$$

and the first formula in (1.25) we get

$$\frac{\partial}{\partial \varepsilon}\rho^{\varepsilon\mu}(z) = \frac{4}{\pi}\iint_U \mu(u) K_{\varepsilon\mu}(z, u) K_{\varepsilon\mu}(u, \bar{z}) d^2u, \tag{6.2}$$

where the integral is understood in the principal value sense. Setting $\varepsilon = 0$ in (6.2) and using Lemma 6.1, we obtain

$$\iint_U \frac{\mu(u)}{(u-z)^2(u-\bar{z})^2} d^2u = 0 \text{ for all } z \in \mathbb{U}. \tag{6.3}$$

Using formulas (1.24) and (1.25), we get from (6.2) the following integral representation for the second variation of the hyperbolic metric

$$L_\mu L_{\bar{\mu}} \rho(z) = -\frac{4}{\pi^2} \iint_U \iint_U \mu(u)\overline{\mu(v)} \left(\frac{1}{(u-\bar{v})^2(z-\bar{v})^2(u-\bar{z})^2} \right.$$
$$\left. + \frac{1}{(u-z)^2(u-\bar{v})^2(\bar{z}-\bar{v})^2} \right) d^2u d^2v. \tag{6.4}$$

The differentiation under the integral sign is justified by the same argument as in [**Ahl62**]. We transform the principal value integrals in (6.4) into the ordinary integrals by using the identity

$$\frac{1}{u-\bar{v}} = \frac{z-\bar{z}}{(u-\bar{z})(z-\bar{v})} + \frac{(\bar{v}-\bar{z})(z-u)}{(u-\bar{z})(z-\bar{v})(u-\bar{v})}, \tag{6.5}$$

which gives

$$\frac{1}{(u-z)^2(u-\bar{v})^2(\bar{z}-\bar{v})^2}$$
$$= \frac{1}{(u-\bar{z})^2(z-\bar{v})^2(u-\bar{v})^2} + \frac{2(z-\bar{z})}{(u-\bar{z})^3(z-\bar{v})^3(u-\bar{v})}$$
$$+ \frac{(z-\bar{z})^2}{(u-z)^2(u-\bar{z})^2(z-\bar{v})^2(\bar{z}-\bar{v})^2} + \frac{2(z-\bar{z})^2}{(u-z)(u-\bar{z})^3(z-\bar{v})^3(\bar{z}-\bar{v})}.$$

Using (6.3) and Corollary 1.15 we see that last two terms in this formula do not contribute into the representation (6.4) and we obtain

$$L_\mu L_{\bar{\mu}} \rho(z) = -\frac{8}{\pi^2} \iint_U \iint_U \mu(u)\overline{\mu(v)} \left(\frac{1}{(u-\bar{v})^2(z-\bar{v})^2(u-\bar{z})^2} \right.$$
$$\left. + \frac{(z-\bar{z})}{(u-\bar{z})^3(u-\bar{v})(z-\bar{v})^3} \right) d^2u d^2v.$$

Now we apply the operator $2(\Delta_0 + \frac{1}{2})$ to the bounded function $\rho^{-1}L_\mu L_{\bar\mu}\rho$ on $\mathbb{U}$. Using (6.5) it is straightforward to compute that

$$(2\Delta_0 + 1)\left(\frac{(z-\bar z)^2}{(u-\bar v)^2(z-\bar v)^2(u-\bar z)^2} + \frac{(z-\bar z)^3}{(u-\bar z)^3(u-\bar v)(z-\bar v)^3}\right)$$
$$= \frac{9}{2}\frac{(z-\bar z)^4}{(u-\bar z)^4(z-\bar v)^4},$$

which, together with (1.23), gives

$$(2\Delta_0 + 1)\left(\rho^{-1}L_\mu L_{\bar\mu}\rho\right)(z)$$
$$= \frac{9}{\pi^2}\iint_\mathbb{U}\iint_\mathbb{U}\mu(u)\overline{\mu(v)}\frac{(z-\bar z)^4}{(u-\bar z)^4(z-\bar v)^4}d^2u\,d^2v = |\mu(z)|^2.$$

Using the property **RK3** in Section 1.4 completes the proof. □

COROLLARY 6.4. *For every $\mu,\nu \in \Omega^{-1,1}(\mathbb{D}^*)$,*

$$G(\mu\bar\nu)(z) = \frac{2}{\pi^2}\iint_\mathbb{U}\iint_\mathbb{U}\mu(u)\overline{\nu(v)}\left(\frac{(z-\bar z)^2}{(u-\bar v)^2(z-\bar v)^2(u-\bar z)^2}\right.$$
$$\left. + \frac{(z-\bar z)^3}{(u-\bar z)^3(u-\bar v)(z-\bar v)^3}\right)d^2u\,d^2v.$$

REMARK 6.5. It follows from Proposition 6.3 by polarization that

$$L_\mu L_{\bar\nu}\rho = \left.\frac{\partial^2}{\partial\varepsilon_1\partial\bar\varepsilon_2}\right|_{\varepsilon_1=\varepsilon_2=0}\rho^{\varepsilon_1\mu+\varepsilon_2\nu} = \rho G(\mu\bar\nu).$$

REMARK 6.6. For the case $\mu \in \Omega^{-1,1}(\mathbb{U},\Gamma)$, where Γ is a cofinite Fuchsian group, the formula for the second variation of the hyperbolic metric in Proposition 6.3 was first proved by Wolpert [**Wol86**]. However, the method in [**Wol86**] does not work for the universal Teichmüller space $T(1)$. The proof of Proposition 6.3 shows that the original singular integral representation (6.4) of Ahlfors can be easily transformed to a closed form using the resolvent kernel.

7. Riemann curvature tensor

In this section we consider $T(1)$ as a Hilbert manifold equipped with the Weil-Petersson metric. We prove that the Weil-Petersson metric is Kähler, compute its Riemann and Ricci tensors, and show that the Ricci, holomorphic, and sectional curvatures are all negative. Since the Weil-Petersson metric is right-invariant, it is sufficient to compute these tensors at the origin of $T(1)$.

7.1. The first variation of the Weil-Petersson metric. For the case $\mu,\nu \in \Omega^{-1,1}(\mathbb{D}^*)$ set

$$Q(\mu,\nu) = P(R(\mu,\nu)) \circ w_\nu \frac{\overline{(w_\nu)_{\bar z}}}{(w_\nu)_z}.$$

PROPOSITION 7.1. *For $\mu,\nu \in \Omega^{-1,1}(\mathbb{D}^*)$,*

$$\left.\frac{\partial}{\partial\varepsilon}Q(\mu,\varepsilon\nu)\right|_{\varepsilon=0} = 0,$$
$$\left.\frac{\partial}{\partial\bar\varepsilon}Q(\mu,\varepsilon\nu)\right|_{\varepsilon=0} = -2\frac{\partial}{\partial\bar z}\rho^{-1}\frac{\partial}{\partial\bar z}G(\mu\bar\nu).$$

7. RIEMANN CURVATURE TENSOR

PROOF. We use complex anti-linear isomorphism $\Omega^{-1,1}(\mathbb{D}^*) \simeq \Omega^{-1,1}(\mathbb{D})$, given by the reflection (1.1), and the model $\mathbb{H}^2 \simeq \mathbb{U}$ of the hyperbolic plane. From (1.6) we get

$$(7.1) \qquad \rho^{\varepsilon\nu}(z)\, Q(\mu, \varepsilon\nu)(z) = \frac{12}{\pi} \iint_{\mathbb{U}} \mu(u) K_{\varepsilon\nu}(u, \bar{z})^2 d^2 u.$$

It follows from equations (1.25) that

$$(7.2) \qquad \left.\frac{\partial}{\partial \varepsilon}\right|_{\varepsilon=0} Q(\mu, \varepsilon\nu)(z) = \frac{6(z - \bar{z})^2}{\pi^2} \iint_{\mathbb{U}} \iint_{\mathbb{U}} \frac{\mu(u)\nu(v)}{(u - \bar{z})^2 (u - v)^2 (v - \bar{z})^2} d^2 u d^2 v$$

and

$$(7.3) \qquad \left.\frac{\partial}{\partial \bar{\varepsilon}}\right|_{\varepsilon=0} Q(\mu, \varepsilon\nu)(z) = \frac{6(z - \bar{z})^2}{\pi^2} \iint_{\mathbb{U}} \iint_{\mathbb{U}} \frac{\mu(u)\overline{\nu(v)}}{(u - \bar{z})^2 (u - \bar{v})^2 (\bar{v} - \bar{z})^2} d^2 u d^2 v.$$

The integrals are understood in the principal value sense and differentiation under the integral sign in (7.1) is justified as in [**Ahl62**].

To prove that the integral (7.2) is zero, we use the identity

$$\frac{1}{(u - v)(u - \bar{z})(v - \bar{z})} = \frac{1}{(u - v)^2}\left(\frac{1}{v - \bar{z}} - \frac{1}{u - \bar{z}}\right),$$

and rewrite the integral (7.2) as follows

$$(7.4) \qquad \iint_{\mathbb{U}} \iint_{\mathbb{U}} \frac{\mu(u)\nu(v)}{(u - \bar{z})^2 (u - v)^2 (v - \bar{z})^2} d^2 u d^2 v$$

$$= \lim_{\varepsilon \to 0} \iiiint_{\mathbb{U} \times \mathbb{U} \setminus \{|u-v| < \varepsilon\}} \left(\frac{1}{(u - v)^4 (v - \bar{z})^2} - \frac{2}{(u - v)^5 (v - \bar{z})}\right.$$

$$\left. + \frac{2}{(u - v)^5 (u - \bar{z})} + \frac{1}{(u - v)^4 (u - \bar{z})^2}\right) \mu(u)\nu(v) d^2 u d^2 v$$

$$= I_1 + I_2 + I_3 + I_4.$$

Applying Lemma (1.12) to the principal value integrals over u in the terms I_1, I_2 and I_4, we conclude that these terms vanish. Changing the order of integrations in I_3 (which is legitimate since domain of integration is invariant under the involution $(u, v) \mapsto (v, u)$) and applying Lemma (1.12) to the integral over v we conclude that the term I_3 also vanishes. This proves that the holomorphic variation of $Q(\mu, \nu)$ vanishes.

To prove the formula for the antiholomorphic variation, we again use the identity (6.5), which gives

$$\frac{1}{(u - \bar{z})^2(u - \bar{v})^2(\bar{v} - \bar{z})^2} = \frac{(u - z)^2}{(u - \bar{z})^4(\bar{v} - z)^2(u - \bar{v})^2}$$

$$+ \frac{(z - \bar{z})^2}{(u - \bar{z})^4(\bar{v} - \bar{z})^2(\bar{v} - z)^2} + \frac{2(z - \bar{z})^2(z - u)}{(u - \bar{z})^5(z - \bar{v})^3(\bar{v} - \bar{z})}$$

$$+ \frac{2(z - \bar{z})(z - u)^2}{(u - \bar{z})^5(z - \bar{v})^3(u - \bar{v})}.$$

Using formula (6.3) and Corollary 1.15 we see that the second and third terms do not contribute to (7.3), and we get

$$(7.5) \quad \frac{\partial}{\partial \bar{\varepsilon}}\bigg|_{\varepsilon=0} Q(\mu, \varepsilon\nu)(z) = \frac{6(z-\bar{z})^2}{\pi^2} \iint_U \iint_U \mu(u)\overline{\nu(v)}$$

$$\left(\frac{(u-z)^2}{(u-\bar{z})^4(\bar{v}-z)^2(u-\bar{v})^2} + \frac{2(z-\bar{z})(z-u)^2}{(u-\bar{z})^5(z-\bar{v})^3(u-\bar{v})} \right) d^2u^2v.$$

Now applying $\partial_{\bar{z}}\rho^{-1}\partial_{\bar{z}}$ to the integral representation in Corollary 6.4 (the differentiation under the integral sign being trivially justified) we get the formula for the antiholomorphic variation. $\square$

Set
$$\dot{Q}(\mu)[\nu] = \frac{\partial}{\partial\bar{\varepsilon}}\bigg|_{\varepsilon=0} Q(\mu, \varepsilon\nu).$$

PROPOSITION 7.2. *Let $\mu \in H^{-1,1}(\mathbb{D}^*)$ and $\nu \in \Omega^{-1,1}(\mathbb{D}^*)$. Then $\dot{Q}(\mu)[\nu] \in L^2(\mathbb{D}^*, \rho(z)d^2z)$ and*

$$\left\|\dot{Q}(\mu)[\nu]\right\|_2^2 = \|\mu\bar{\nu}\|_2^2 - (\mu\bar{\nu}, G(\mu\bar{\nu})),$$

where $(\ ,\ )$ stands for the inner product in the Hilbert space $L^2(\mathbb{D}^*, \rho(z)d^2z)$.

PROOF. As in the proof of Proposition 7.1, it is convenient to use the isomorphism $\Omega^{-1,1}(\mathbb{D}^*) \simeq \Omega^{-1,1}(\mathbb{D})$. For $\mu \in BC^\infty(\mathbb{D}) \cap L^2(\mathbb{D}, \rho(z)d^2z)$ with compact support and $\nu \in BC^\infty(\mathbb{D})$ we set

$$\mathcal{Q}_\nu(\mu) = 2\frac{\partial}{\partial z}\rho^{-1}\frac{\partial}{\partial \bar{z}}G(\mu\bar{\nu}).$$

We will prove that $\|\mathcal{Q}_\nu(\mu)\|_2^2 = \|\mu\bar{\nu}\|_2^2 - (\mu\bar{\nu}, G(\mu\bar{\nu}))$, so that $\mathcal{Q}_\nu$ extends to a bounded linear operator on $L^2(\mathbb{D}, \rho(z)d^2z)$. Since, according to Proposition 7.1, $\mathcal{Q}_\nu(\mu) = -\dot{Q}(\mu)[\nu]$ for $\mu \in H^{-1,1}(\mathbb{D}^*)$ and $\nu \in \Omega^{-1,1}(\mathbb{D}^*)$, the assertion follows from this fact.

From the explicit formula (1.19) we get the following estimates

$$(7.6) \qquad G(z,w) = O((1-|z|^2)), \quad (\partial_z\rho^{-1}\partial_z)G(z,w) = O((1-|z|^2)),$$
$$\partial_z G(z,w) = O(1), \quad \left(\partial_{\bar{z}}\left(\partial_z\rho^{-1}\partial_z\right)\right)G(z,w) = O(1) \quad \text{as} \quad |z| \to 1,$$

uniformly as w varies on compact subsets of $\mathbb{D}$. Using Stokes' theorem and the identity

$$\rho^{-1}\partial_z\,\rho^{-1}\partial_z\,\rho\,\partial_{\bar{z}}\,\rho^{-1}\partial_{\bar{z}} = \Delta_0(\Delta_0 + \tfrac{1}{2})$$

we get

$$\iint_{\mathbb{D}} |\mathcal{Q}_\nu(\mu)|^2 \rho(z)d^2z = 4\iint_{\mathbb{D}} \partial_z\left(\rho^{-1}\partial_z G(\bar{\mu}\nu)\right)\partial_{\bar{z}}\left(\rho^{-1}\partial_{\bar{z}}G(\mu\bar{\nu})\right)\rho(z)d^2z$$

$$= 4\iint_{\mathbb{D}} \Delta_0(\Delta_0 + \tfrac{1}{2})G(\mu\bar{\nu})G(\bar{\mu}\nu)\rho(z)d^2z$$

$$= 2\iint_{\mathbb{D}} \Delta_0(\mu\bar{\nu})G(\bar{\mu}\nu)\rho(z)d^2z,$$

where in the last line we have used property **RK3** from Section 1.4. Due to the estimates (7.6) the boundary terms arising in the Stokes' formula vanish. Using Stokes' theorem once again we finally get

$$\iint_{\mathbb{D}} |\mathcal{Q}_\nu(\mu)|^2 \rho(z) d^2 z = 2 \iint_{\mathbb{D}} \mu\bar{\nu}\Delta_0 G(\bar{\mu}\nu)\rho(z) d^2 z$$
$$= \iint_{\mathbb{D}} |\mu\bar{\nu}|^2 \rho(z) d^2 z - \iint_{\mathbb{D}} \mu\bar{\nu} G(\bar{\mu}\nu)\rho(z) d^2 z$$
$$= \|\mu\bar{\nu}\|_2^2 - (\mu\bar{\nu}, G(\mu\bar{\nu})).$$

The boundary terms again vanish due to (7.6) and Remark 3.2. □

COROLLARY 7.3. *For $\mu, \nu \in H^{-1,1}(\mathbb{D}^*)$ and $\kappa \in \Omega^{-1,1}(\mathbb{D}^*)$,*

$$\left(\dot{Q}(\mu)[\kappa], \dot{Q}(\nu)[\kappa]\right) = (\mu\bar{\kappa}, \nu\bar{\kappa}) - (\mu\bar{\kappa}, G(\nu\bar{\kappa})).$$

THEOREM 7.4. *For $\mu, \nu \in H^{-1,1}(\mathbb{D}^*)$ and $\kappa \in \Omega^{-1,1}(\mathbb{D}^*)$,*

$$\left.\frac{\partial}{\partial\varepsilon} g_{\mu\bar{\nu}}(\varepsilon\kappa)\right|_{\varepsilon=0} = 0.$$

PROOF. Since $P^2 = P$, we get from (4.3),
(7.7)
$$g_{\mu\bar{\nu}}(\kappa) = \iint_{\mathbb{D}^*} Q(\mu,\kappa)\overline{Q(\nu,\kappa)}(1-|\kappa|^2)w_\kappa^*(\rho)(z) d^2 z = \iint_{\mathbb{D}^*} \mu\overline{Q(\nu,\kappa)} w_\kappa^*(\rho)(z) d^2 z,$$

so that

$$\left.\frac{\partial}{\partial\varepsilon} g_{\mu\bar{\nu}}(\varepsilon\kappa)\right|_{\varepsilon=0} = \left(\dot{Q}(\mu)[\kappa], \nu\right) + \left(\mu, \dot{Q}(\nu)[\kappa]\right) = \left(\mu, \dot{Q}(\nu)[\kappa]\right).$$

Differentiation under the integral sign is justified as in [**Ahl62**]. Thus for all $\mu, \nu \in H^{-1,1}(\mathbb{D}^*)$ and $\kappa \in \Omega^{-1,1}(\mathbb{D}^*)$

$$\left(\dot{Q}(\mu)[\kappa], \nu\right) = 0,$$

and the theorem follows. □

Let $\{\mu_n\}_{n=2}^\infty$ be an orthonormal basis for the Hilbert space $H^{-1,1}(\mathbb{D}^*)$,

$$\mu_n(z) = -\sqrt{\frac{n^3-n}{8\pi}}(1-|z|^2)^2 \bar{z}^{-n-2}, \quad n = 2, 3, \ldots,$$

and let $\{\varepsilon_n\}_{n=2}^\infty$ be the corresponding Bers coordinates on the chart V_0. Since $\|\mu\|_2 = 2\|D_0\beta(\mu)\|_2$, it follows from Section 2.3 that $\sum_{n=2}^\infty |\varepsilon_n|^2 < \frac{4\pi}{3}$. Denote by $\frac{\partial}{\partial\varepsilon_n}$ the corresponding directional derivatives — the vector field $\frac{\partial}{\partial\varepsilon_{\mu_n}}$ on V_0, and set $g_{m\bar{n}} = g_{\mu_m\bar{\mu}_n}$. Since the basis $\{\mu_n\}_{n=2}^\infty$ is orthonormal, $g_{m\bar{n}} = \delta_{mn}$ at the origin of $T(1)$.

COROLLARY 7.5. *The Weil-Petersson metric is a Kähler metric on the Hilbert manifold $T(1)$, and the Bers coordinates are geodesic coordinates at the origin of $T(1)$.*

PROOF. It follows from Theorem 7.4 that
$$\frac{\partial g_{m\bar{n}}}{\partial \varepsilon_l}(0) = 0.$$

□

REMARK 7.6. Propositions 7.1 and 7.2 and Theorem 7.4 generalize Wolpert's results for finite-dimensional Teichmüller spaces (see Lemma 2.7 and Theorem 2.9 in [**Wol86**]) to the universal Teichmüller space. In particular, our proof of Theorem 7.4 (after Proposition 7.2 has been established) is the same as in [**Wol86**].

7.2. The second variation of the Weil-Petersson metric. Due to Corollary 7.5, the Riemann tensor of the Weil-Petersson metric at the origin of $T(1)$ is given by
$$R_{k\bar{l}m\bar{n}} = -\frac{\partial^2 g_{k\bar{l}}}{\partial \varepsilon_m \partial \bar{\varepsilon}_n}(0),$$
where we are using conventions of Yano and Bochner [**YB53**] in Hermitian geometry.

THEOREM 7.7. For $\mu, \nu \in H^{-1,1}(\mathbb{D}^*)$ and $\kappa \in \Omega^{-1,1}(\mathbb{D}^*)$,
$$\left.\frac{\partial^2}{\partial \varepsilon \partial \bar{\varepsilon}} g_{\mu\bar{\nu}}(\varepsilon\kappa)\right|_{\varepsilon=0} = (\mu\bar{\kappa}, G(\nu\bar{\kappa})) + (\mu\bar{\nu}, G(|\kappa|^2)).$$

PROOF. Differentiating the representation (7.7) for $g_{\mu\bar{\nu}}(\varepsilon\kappa)$ with respect to ε and $\bar{\varepsilon}$ we get

$$\left.\frac{\partial^2}{\partial \varepsilon \partial \bar{\varepsilon}} g_{\mu\bar{\nu}}(\varepsilon\kappa)\right|_{\varepsilon=0}$$
$$= \left(\left.\frac{\partial^2}{\partial \varepsilon \partial \bar{\varepsilon}} Q(\mu, \varepsilon\kappa)\right|_{\varepsilon=0}, \nu\right) + \left(\mu\bar{\nu}, \rho^{-1} \left.\frac{\partial^2}{\partial \varepsilon \partial \bar{\varepsilon}} \rho^{\varepsilon\kappa}\right|_{\varepsilon=0}\right)$$
$$= \left(\left.\frac{\partial^2}{\partial \varepsilon \partial \bar{\varepsilon}} Q(\mu, \varepsilon\kappa)\right|_{\varepsilon=0}, \nu\right) + \left(\left.\frac{\partial}{\partial \bar{\varepsilon}}\right|_{\varepsilon=0} Q(\mu, \varepsilon\kappa), \left.\frac{\partial}{\partial \varepsilon}\right|_{\varepsilon=0} Q(\nu, \varepsilon\kappa)\right)$$
$$+ \left(\mu, \left.\frac{\partial^2}{\partial \varepsilon \partial \bar{\varepsilon}} Q(\nu, \varepsilon\kappa)\right|_{\varepsilon=0}\right) + \left(\mu\bar{\nu}, \rho^{-1} \left.\frac{\partial^2}{\partial \varepsilon \partial \bar{\varepsilon}} \rho^{\varepsilon\kappa}\right|_{\varepsilon=0}\right) - (\mu\bar{\nu}, |\kappa|^2).$$

The differentiation under the integral sign is justified as in [**Ahl62**], provided that all integrals above are absolutely convergent. This follows from Proposition 6.3, property **RK3** in Section 1.4, Proposition 7.2 and the following

LEMMA 7.8. For $\mu \in H^{-1,1}(\mathbb{D}^*)$ and $\nu \in \Omega^{-1,1}(\mathbb{D}^*)$,
$$\left.\frac{\partial^2}{\partial \varepsilon \partial \bar{\varepsilon}} Q(\mu, \varepsilon\nu)\right|_{\varepsilon=0} \in L^2(\mathbb{D}^*, \rho(z)d^2z).$$

PROOF. Here we prove Lemma 7.8. Using the model $\mathbb{H} \simeq \mathbb{U}$ and (1.25), we get for the second variation of Q,

$$\left.\frac{\partial^2}{\partial \varepsilon \partial \bar{\varepsilon}}\right|_{\varepsilon=0} Q(\mu, \varepsilon\nu)(z)$$

$$= \frac{24}{\pi^3} \iint_{\mathbb{U}} \iint_{\mathbb{U}} \iint_{\mathbb{U}} \frac{\rho(z)^{-1}\mu(w)\overline{\nu(u)}\nu(v)d^2v d^2u d^2w}{(w-v)^2(v-\bar{z})^2(w-\bar{u})^2(\bar{u}-\bar{z})^2}$$

$$+ \frac{24}{\pi^3} \iint_{\mathbb{U}} \iint_{\mathbb{U}} \iint_{\mathbb{U}} \frac{\rho(z)^{-1}\mu(w)\overline{\nu(u)}\nu(v)d^2v d^2u d^2w}{(w-\bar{z})^2(w-v)^2(v-\bar{u})^2(\bar{u}-\bar{z})^2}$$

$$+ \frac{24}{\pi^3} \iint_{\mathbb{U}} \iint_{\mathbb{U}} \iint_{\mathbb{U}} \frac{\rho(z)^{-1}\mu(w)\overline{\nu(u)}\nu(v)d^2v d^2u d^2w}{(w-\bar{z})^2(w-\bar{u})^2(v-\bar{u})^2(v-\bar{z})^2}$$

$$- \mu(z)\rho^{-1}(z) \left.\frac{\partial^2}{\partial \varepsilon \partial \bar{\varepsilon}}\right|_{\varepsilon=0} \rho^{\varepsilon\nu}(z) = I_1(z) + I_2(z) + I_3(z) + I_4(z).$$

Using Proposition 6.3 and property **RK3**, we get that $I_4 \in L^2(\mathbb{U}, \rho(z)d^2z)$. Now using the Cauchy-Schwarz inequality, the identity

$$\iint_{\mathbb{U}} \frac{d^2w}{|w-\bar{z}|^4} = \frac{\pi}{4}\rho(z),$$

and the property that the Hilbert transform is an isometry on $L^2(\mathbb{C}, d^2z)$, we get

$$\|I_1\|_2^2 = \frac{24^2}{\pi^6} \iint_{\mathbb{U}} \left| \iint_{\mathbb{U}} \iint_{\mathbb{U}} \iint_{\mathbb{U}} \frac{\mu(w)\overline{\nu(u)}\nu(v)d^2v d^2u d^2w}{(w-v)^2(v-\bar{z})^2(w-\bar{u})^2(\bar{u}-\bar{z})^2} \right|^2 \rho(z)^{-1} d^2z$$

$$\leq \frac{24^2}{\pi^6} \iint_{\mathbb{U}} \left(\iint_{\mathbb{U}} \frac{\rho(z)^{-1}|\nu(v)|^2 d^2v}{|v-\bar{z}|^4} \right)$$

$$\times \left(\iint_{\mathbb{U}} \left| \iint_{\mathbb{U}} \iint_{\mathbb{U}} \frac{\mu(w)\overline{\nu(u)}d^2u d^2w}{(w-v)^2(w-\bar{u})^2(\bar{u}-\bar{z})^2} \right|^2 d^2v \right) d^2z$$

$$\leq \frac{12^2 \|\nu\|_\infty^2}{\pi^5} \iint_{\mathbb{U}} \iint_{\mathbb{U}} \left| \iint_{\mathbb{U}} \iint_{\mathbb{U}} \frac{\mu(w)\overline{\nu(u)}d^2u d^2w}{(w-v)^2(w-\bar{u})^2(\bar{u}-\bar{z})^2} \right|^2 d^2v d^2z$$

$$\leq \frac{12^2 \|\nu\|_\infty^2}{\pi^3} \iint_{\mathbb{U}} \iint_{\mathbb{U}} |\mu(v)|^2 \left| \iint_{\mathbb{U}} \frac{\nu(u)d^2u}{(u-\bar{v})^2(u-z)^2} \right|^2 d^2z d^2v \leq 36\|\nu\|_\infty^4 \|\mu\|_2^2.$$

Similarly, denoting by $\overline{\mathbb{U}}$ the lower half-plane,

$$\|I_3\|_2^2 = \frac{24^2}{\pi^6} \iint\limits_{\mathbb{U}} \left| \iint\limits_{\mathbb{U}} \iint\limits_{\mathbb{U}} \iint\limits_{\mathbb{U}} \frac{\mu(w)\overline{\nu(u)}\nu(v)d^2vd^2ud^2w}{(w-\bar{z})^2(w-\bar{u})^2(v-\bar{u})^2(v-\bar{z})^2} \right|^2 \rho(z)^{-1}d^2z$$

$$\leq \frac{12^2\|\nu\|_\infty^2}{\pi^5} \iint\limits_{\mathbb{U}} \iint\limits_{\mathbb{U}} \left| \iint\limits_{\mathbb{U}} \iint\limits_{\mathbb{U}} \frac{\mu(w)\overline{\nu(u)}d^2ud^2w}{(w-\bar{z})^2(w-\bar{u})^2(v-\bar{u})^2} \right|^2 d^2vd^2z$$

$$= \frac{12^2\|\nu\|_\infty^2}{\pi^5} \iint\limits_{\mathbb{U}} \iint\limits_{\overline{\mathbb{U}}} \left| \iint\limits_{\mathbb{U}} \iint\limits_{\mathbb{U}} \frac{\mu(w)\overline{\nu(u)}d^2ud^2w}{(w-\bar{z})^2(w-\bar{u})^2(\bar{v}-\bar{u})^2} \right|^2 d^2vd^2z$$

$$\leq \frac{12^2\|\nu\|_\infty^2}{\pi^3} \iint\limits_{\mathbb{U}} \iint\limits_{\mathbb{U}} |\nu(v)|^2 \left| \iint\limits_{\mathbb{U}} \frac{\mu(w)d^2w}{(w-\bar{z})^2(w-\bar{v})^2} \right|^2 d^2vd^2z$$

$$\leq \frac{12^2\|\nu\|_\infty^4}{\pi} \iint\limits_{\mathbb{U}} \iint\limits_{\mathbb{U}} \frac{|\mu(v)|^2}{|v-\bar{z}|^4} d^2vd^2z = 36\|\nu\|_\infty^4\|\mu\|_2^2.$$

For the term I_2 we use the same identity as in the proof of Proposition 6.3[5]

$$\frac{1}{(v-\bar{u})^2(\bar{u}-\bar{z})^2} = \frac{(v-z)^2}{(v-\bar{z})^2(z-\bar{u})^2(v-\bar{u})^2} + \frac{2(v-z)^2(z-\bar{z})}{(v-\bar{z})^3(v-\bar{u})(z-\bar{u})^3}$$
$$+ \frac{(z-\bar{z})^2}{(v-\bar{z})^2(z-\bar{u})^2(\bar{u}-\bar{z})^2} + \frac{2(z-\bar{z})^2(v-z)}{(v-\bar{z})^3(\bar{u}-\bar{z})(\bar{u}-z)^3}.$$

As in the proof of Proposition 6.3, the last two terms do not contribute to I_2, and we obtain

$$I_2(z) = \frac{24}{\pi^3}\rho^{-1}(z) \iint\limits_{\mathbb{U}} \iint\limits_{\mathbb{U}} \iint\limits_{\mathbb{U}} \frac{\mu(w)\nu(v)\overline{\nu(u)}(v-z)^2d^2vd^2ud^2w}{(w-\bar{z})^2(w-v)^2(v-\bar{z})^2(z-\bar{u})^2(v-\bar{u})^2}$$
$$+ \frac{48}{\pi^3}\rho^{-1}(z) \iint\limits_{\mathbb{U}} \iint\limits_{\mathbb{U}} \iint\limits_{\mathbb{U}} \frac{\mu(w)\nu(v)\overline{\nu(u)}(v-z)^2(z-\bar{z})d^2vd^2ud^2w}{(w-\bar{z})^2(w-v)^2(v-\bar{z})^3(z-\bar{u})^3(v-\bar{u})}$$
$$= I_5(z) + I_6(z).$$

---

[5]We could estimate I_2 in the same way as I_1 If $\nu \in H^{-1,1}(\mathbb{U})$. However, for Theorem 8.5 we only have $\nu \in \Omega^{-1,1}(\mathbb{U})$.

The L^2-norm of I_5 is estimated exactly as before and we get $\|I_5\|_2^2 \leq 36\|\nu\|_\infty^4\|\mu\|_2^2$.
Finally,

$$\|I_6\|_2^2 = \frac{48^2}{\pi^6}\iint_U \left|\iint_U \iint_U \iint_U \frac{\mu(w)\nu(v)\overline{\nu(u)}(v-z)^2(z-\bar{z})d^2vd^2ud^2w}{(w-\bar{z})^2(w-v)^2(v-\bar{z})^3(z-\bar{u})^3(v-\bar{u})}\right|^2 \rho(z)^{-1}d^2z$$

$$\leq \frac{24^2\|\nu\|_\infty^2}{\pi^5}\iiiint_{U\ U}\frac{|z-\bar{z}|^2}{|u-\bar{z}|^2}\left|\iiiint_{U\ U}\frac{\mu(w)\nu(v)(v-z)^2d^2vd^2w}{(w-\bar{z})^2(w-v)^2(v-\bar{z})^3(v-\bar{u})}\right|^2 d^2ud^2z$$

$$\leq \frac{24^2\|\nu\|_\infty^2}{\pi^5}\iint_U\left(\iint_U\left|\iint_U\frac{\mu(w)d^2w}{(w-\bar{z})^2(w-v)^2}\right|^2 d^2v\right)$$

$$\left(\iint_U\frac{|z-\bar{z}|^2}{|u-\bar{z}|^2}\iint_U\frac{|\nu(v)|^2|v-z|^4}{|v-\bar{z}|^6|v-\bar{u}|^2}d^2vd^2u\right)d^2z.$$

Making a change of variables $\frac{v-z}{v-\bar{z}} \mapsto v$ and $\frac{u-z}{u-\bar{z}} \mapsto u$, we obtain

$$\iint_U\frac{|z-\bar{z}|^2}{|u-\bar{z}|^2}\iint_U\frac{|\nu(v)|^2|v-z|^4}{|v-\bar{z}|^6|v-\bar{u}|^2}d^2vd^2u \leq \|\nu\|_\infty^2\iint_D\iint_D\frac{|v|^4}{|1-v\bar{u}|^2}d^2vd^2u$$

$$= \frac{3\pi^2}{4}\|\nu\|_\infty^2.$$

Hence

$$\|I_6\|_2^2 \leq \frac{3\cdot 12^2\|\nu\|_\infty^4}{\pi}\iint_U\iint_U\frac{|\mu(v)|^2}{|v-\bar{z}|^4}d^2vd^2z = 108\|\nu\|_\infty^4\|\mu\|_2^2.$$

$\square$

Now comparing the two expressions for the second variation of $g_{\mu\bar{\nu}}$ and using Corollary 7.3 we get

$$\left(\mu, \frac{\partial^2}{\partial\varepsilon\partial\bar{\varepsilon}}Q(\nu,\varepsilon\kappa)\bigg|_{\varepsilon=0}\right) = -\left(\frac{\partial}{\partial\bar{\varepsilon}}\bigg|_{\varepsilon=0}Q(\mu,\varepsilon\kappa), \frac{\partial}{\partial\bar{\varepsilon}}\bigg|_{\varepsilon=0}Q(\nu,\varepsilon\kappa)\right) + (\mu\bar{\nu},|\kappa|^2)$$
$$= (\mu\bar{\kappa}, G(\nu\bar{\kappa})).$$

Using Proposition 6.3, we finally obtain

$$\frac{\partial^2}{\partial\varepsilon\partial\bar{\varepsilon}}\bigg|_{\varepsilon=0}g_{\mu\bar{\nu}}(\varepsilon\kappa) = (\mu\bar{\kappa}, G(\nu\bar{\kappa})) + (\mu\bar{\nu}, G(|\kappa|^2)).$$

$\square$

COROLLARY 7.9. *At the origin of $T(1)$,*

$$R_{k\bar{l}m\bar{n}} = -(\mu_k\bar{\mu}_l, G(\bar{\mu}_m\mu_n)) - (\bar{\mu}_l\mu_m, G(\bar{\mu}_k\mu_n)).$$

PROOF. It follows from Theorem 7.7 by polarization that

$$R_{\kappa\bar{\lambda}\mu\bar{\nu}} = -\frac{\partial^2}{\partial\varepsilon_1\partial\bar{\varepsilon}_2}\bigg|_{\varepsilon_1=\varepsilon_2=0}g_{\mu\bar{\nu}}(\varepsilon_1\kappa+\varepsilon_2\lambda)$$
$$= -(\kappa\bar{\lambda}, G(\bar{\mu}\nu)) - (\bar{\lambda}\mu, G(\bar{\kappa}\nu)).$$

$\square$

REMARK 7.10. For finite-dimensional Teichmüller spaces this result was proved by Wolpert [**Wol86**]. Except Lemma 7.8, our derivation is the same as in [**Wol86**].

7.3. Ricci and sectional curvatures. The Ricci tensor at the origin of $T(1)$ for the orthonormal basis $\{\mu_n\}_{n=2}^\infty$ of $H^{-1,1}(\mathbb{D}^*)$ is defined by the following series

$$\mathcal{R}_{k\bar{l}} = \sum_{n=2}^\infty R_{k\bar{n}n\bar{l}}.$$

THEOREM 7.11. *The Ricci tensor at the origin of $T(1)$ is well-defined and is given by*

$$\mathcal{R}_{k\bar{l}} = -\frac{13}{12\pi}\delta_{kl}.$$

PROOF. Set $\mu = \mu_k, \nu = \mu_l$, and $\mathcal{R}_{\mu\bar{\nu}} = \mathcal{R}_{k\bar{l}}$. We have

$$\mathcal{R}_{\mu\bar{\nu}} = -\frac{2}{\pi}\sum_{n=2}^\infty (n^3-n) \left(\iint_{\mathbb{D}^*}\iint_{\mathbb{D}^*} G(z,w)\mu(z)z^{n-2}\overline{\nu(w)}\bar{w}^{n-2}d^2wd^2z \right.$$

$$\left. + \iint_{\mathbb{D}^*}\iint_{\mathbb{D}^*} G(z,w)\bar{z}^{n-2}z^{n-2}\mu(w)\overline{\nu(w)}\frac{(1-|z|^2)^2}{(1-|w|^2)^2}d^2wd^2z \right)$$

$$= -\frac{12}{\pi}\iint_{\mathbb{D}^*}\iint_{\mathbb{D}^*} G(z,w)\frac{\mu(z)\overline{\nu(w)}}{(1-z\bar{w})^4}d^2wd^2z$$

$$-\frac{3}{4\pi}\iint_{\mathbb{D}^*}\iint_{\mathbb{D}^*} G(z,w)\mu(w)\overline{\nu(w)}\rho(z)\rho(w)d^2wd^2z$$

$$= I_1 + I_2.$$

For the second integral, we use property **RK4** in Section 1.4 and get

$$I_2 = -\frac{3}{4\pi}\iint_\mathbb{D} \mu(w)\overline{\nu(w)}\rho(w)d^2w = -\frac{3}{4\pi}g_{\mu\bar{\nu}}.$$

For the first integral, we use projection formula (1.7) and get

$$I_1 = -\frac{36}{\pi^2}\iint_{\mathbb{D}^*}\iint_{\mathbb{D}^*}\iint_{\mathbb{D}^*} G(z,w)\mu(z)\overline{\nu(v)}\frac{(1-|w|^2)^2}{(1-z\bar{w})^4(1-w\bar{v})^4}d^2vd^2wd^2z.$$

Let

$$B(z,v) = \iint_{\mathbb{D}^*} G(z,w)\frac{(1-|w|^2)^2}{(1-w\bar{v})^4(1-z\bar{w})^4}d^2w.$$

The kernel $B(z,v)$ satisfies

(7.8) $\qquad B(z,v) = B\Big(\sigma z, \sigma v\Big)\sigma'(z)^2\overline{\sigma'(v)}^2 \quad \text{for all} \quad \sigma \in \text{PSU}(1,1),$

and

$$B(0,v) = \iint_{\mathbb{D}^*}\left(\frac{1}{2\pi}\frac{1+|w|^2}{1-|w|^2}\log\frac{1}{|w|^2} - \frac{1}{\pi}\right)\frac{(1-|w|^2)^2}{(1-\bar{v}w)^4}d^2w = \frac{1}{9}.$$

Hence
$$B(z,v) = \frac{1}{9(1-z\bar{v})^4}$$
and
$$I_1 = -\frac{4}{\pi^2} \iint_{\mathbb{D}^*} \iint_{\mathbb{D}^*} \mu(z)\overline{\nu(v)} \frac{1}{(1-z\bar{v})^4} d^2z\, d^2w$$
$$= -\frac{1}{3\pi} \iint_{\mathbb{D}^*} \mu(z)\overline{\nu(z)}\rho(z) d^2z = -\frac{1}{3\pi} g_{\mu\bar{\nu}}.$$

Therefore,
$$\mathcal{R}_{\mu\bar{\nu}} = -\left(\frac{3}{4\pi} + \frac{1}{3\pi}\right) g_{\mu\bar{\nu}} = -\frac{13}{12\pi} g_{\mu\bar{\nu}}.$$

□

Since the Weil-Petersson metric on $T(1)$ is right-invariant, it follows from Theorem 7.11 that the Ricci tensor is well-defined everywhere on $T(1)$. Denote by Ric_{WP} corresponding Ricci $(1,1)$-form on $T(1)$. In terms of Bers coordinates $\{\varepsilon_n\}_{n=2}^\infty$ on the coordinate chart V_μ the Ricci form is given by

$$Ric_{WP} = \frac{i}{2} \sum_{k,l=2}^\infty \mathcal{R}_{k\bar{l}}\, d\varepsilon_k \wedge d\bar{\varepsilon}_l.$$

COROLLARY 7.12. *The universal Teichmüller space $T(1)$ is a Kähler-Einstein manifold with negative constant Ricci curvature,*

$$Ric_{WP} = -\frac{13}{12\pi} \omega_{WP}.$$

PROOF. Since $(1,1)$-forms ω_{WP} and Ric are right-invariant, the result immediately follows from Theorem 7.11 □

REMARK 7.13. For the dense submanifold $\text{Möb}(S^1)\backslash \text{Diff}_+(S^1)$ of $T_0(1)$ the statement of Theorem 7.11 was established by different methods in [**KY87**] and [**BR87a, BR87b**]. The "magic ratio" $\frac{13}{12\pi}$ is omnipresent in mathematics and in the string theory.

Let $\frac{\partial}{\partial t_\mu}, \frac{\partial}{\partial t_\nu} \in T_0^\mathbb{R} T(1)$ be real tangent vectors. According to [**YB53**], the sectional curvature of the section spanned by these vectors is given by R/g, where

(7.9) $$R = R_{\mu\bar{\nu}\nu\bar{\mu}} + R_{\nu\bar{\mu}\mu\bar{\nu}} - R_{\mu\bar{\nu}\mu\bar{\nu}} - R_{\nu\bar{\mu}\nu\bar{\mu}},$$
$$g = 4g_{\mu\bar{\mu}}g_{\nu\bar{\nu}} - 2|g_{\mu\bar{\nu}}|^2 - 2\operatorname{Re}(g_{\mu\bar{\nu}})^2.$$

Similarly, the holomorphic sectional curvature of the section spanned by the holomorphic tangent vector $\frac{\partial}{\partial \varepsilon_\mu}$, where $g_{\mu\bar{\mu}} = 1$, is given by $R_{\mu\bar{\mu}\mu\bar{\mu}}$.

As in the finite-dimensional case [**Wol86**], we have

THEOREM 7.14. *The sectional and holomorphic sectional curvatures of $T(1)$ are negative.*

PROOF. For a section spanned by $\frac{\partial}{\partial\varepsilon_\mu}$, the holomorphic sectional curvature is obviously negative: $\mu \neq 0$ so that $G(|\mu|^2) > 0$, and $(|\mu|^2, G(|\mu|^2)) > 0$.

For a section spanned by the real tangent vectors $\frac{\partial}{\partial t_\mu}$ and $\frac{\partial}{\partial t_\nu}$, using Cauchy-Schwarz inequality, it is easy to see that g is positive. Using Corollary 7.9, the properties **RK1** and **RK2**, we get

$$R = 4\mathrm{Re}\,(\mu\bar{\nu}, G(\bar{\mu}\nu)) - 2(\mu\bar{\nu}, G(\mu\bar{\nu})) - 2(|\mu|^2, G(|\nu|^2)).$$

From the property **RK2** and Cauchy-Schwarz inequality we have

$$|G(\mu\bar{\nu})(z)| \leq \iint_{\mathbb{D}^*} G(z,w)^{1/2}|\mu(w)|G(z,w)^{1/2}|\nu(w)|\rho(w)d^2w$$

$$\leq \left(\iint_{\mathbb{D}^*} G(z,w)|\mu(w)|^2\rho(w)d^2w\right)^{1/2} \left(\iint_{\mathbb{D}^*} G(z,w)|\nu(w)|^2\rho(w)d^2w\right)^{1/2}$$

so that

$$|(\bar{\mu}\nu, G(\mu\bar{\nu}))| \leq \iint_{\mathbb{D}^*} |\mu\nu|G(|\mu|^2)^{1/2}G(|\nu|^2)^{1/2}\rho(z)d^2z$$

$$\leq \left(\iint_{\mathbb{D}^*} |\mu|^2 G(|\nu|^2)\rho(z)d^2z\right)^{1/2} \left(\iint_{\mathbb{D}^*} |\nu|^2 G(|\mu|^2)\rho(z)d^2z\right)^{1/2}$$

$$= (|\mu|^2, G(|\nu|^2)).$$

On the other hand, if we let $\mu\bar{\nu} = \alpha + i\beta$, where α and β are real-valued functions, then $\mathrm{Re}\,(\mu\bar{\nu}, G(\bar{\mu}\nu)) = (\alpha, G(\alpha)) - (\beta, G(\beta))$ and $(\mu\bar{\nu}, G(\mu\bar{\nu})) = (\alpha, G(\alpha)) + (\beta, G(\beta))$. Since for a bounded real-valued smooth function h, we can deduce from the property **RK3** and the positivity of the operator $\Delta_0 + \frac{1}{2}$ that $(h, G(h)) \geq 0$, it follows that $\mathrm{Re}\,(\mu\bar{\nu}, G(\bar{\mu}\nu)) \leq (\mu\bar{\nu}, G(\mu\bar{\nu}))$. Hence R is negative by Cauchy-Schwarz inequality. $\square$

8. Finite-dimensional Teichmüller spaces

Curvature properties of finite-dimensional Teichmüller spaces were extensively studied by Ahlfors [**Ahl62**], Royden [**Roy75**], and especially by Wolpert [**Wol86**]. Here, for the Teichmüller space $T(\Gamma)$ of a cofinite Fuchsian group Γ we show how to get Wolpert's explicit formulas from the curvature formulas of the Hilbert manifold $T(1)$, derived in Section 7.

First note that the canonical embedding of the finite-dimensional complex manifold $T(\Gamma)$ into $T(1)$ is holomorphic with respect to the Banach manifold structure on $T(1)$ but not with respect to the Hilbert manifold structure on $T(1)$. Indeed, for a cofinite Fuchsian group Γ the finite-dimensional vector space $\Omega^{-1,1}(\mathbb{D}^*, \Gamma)$ is not a subspace of the Hilbert space $H^{-1,1}(\mathbb{D}^*)$, but rather

$$\Omega^{-1,1}(\mathbb{D}^*, \Gamma) \cap H^{-1,1}(\mathbb{D}^*) = \{0\}.$$

Thus the Weil-Petersson metric on $T(\Gamma)$, defined in Section 1.3, is not a pullback of the Weil-Petersson metric on $T(1)$. However, due to Lemma 1.9 we can represent the Petersson inner product on the tangent space at the origin of $T(\Gamma)$ as an "average" of inner products in $T(1)$. Namely, using the canonical complex anti-linear isomorphisms $\Omega^{-1,1}(\mathbb{D}^*,\Gamma) \simeq \Omega^{-1,1}(\mathbb{D},\Gamma)$ and $H^{-1,1}(\mathbb{D}^*) \simeq H^{-1,1}(\mathbb{D})$, we have

$$\langle \mu,\nu\rangle_{WP} = \iint_{\Gamma\backslash\mathbb{D}} \mu\bar{\nu}\rho(z)d^2z = \lim_{r\to 1^-} \frac{A(\Gamma\backslash\mathbb{D})}{A(\mathbb{D}_r)} \iint_{\mathbb{D}_r} \mu\bar{\nu}\rho(z)d^2z$$

$$= \lim_{r\to 1^-} \frac{A(\Gamma\backslash\mathbb{D})}{A(\mathbb{D}_r)} \iint_{\mathbb{D}} \mu_r\bar{\nu}_r\rho(z)d^2z.$$

Here $\mu,\nu \in \Omega^{-1,1}(\mathbb{D},\Gamma)$ and $\mu_r = \chi_r\mu, \nu_r = \chi_r\nu$, where χ_r is the characteristic function of $\mathbb{D}_r = \{z \in \mathbb{D} : |z| \leq r\}$. In what follows we will denote by $(\ ,\ )_\Gamma$ the Petersson inner product $\langle\ ,\ \rangle_{WP}$ in $\Omega^{-1,1}(\mathbb{D},\Gamma)$, as well as the inner product for the Hilbert space $L^2(\Gamma\backslash\mathbb{D}, \rho(z)d^2z)$. Since they are given by the same formula, there would be no confusion. Moreover, for $\mu \in \Omega^{-1,1}(\mathbb{D},\Gamma)$, $|\mu| \in L^2(\Gamma\backslash\mathbb{D}, \rho(z)d^2z)$.

LEMMA 8.1. *Let $\mu \in \Omega^{-1,1}(\mathbb{D})$ and $\nu \in L^\infty(\mathbb{D}) \cap L^1(\mathbb{D},\rho(z)d^2z)$. Then*

(i) *For $0 < r < 1$,*

$$P(\mu_r) \in H^{-1,1}(\mathbb{D}) \quad \text{and} \quad P(\mu_r)(z) = O\left((1-|z|^2)^2\right) \quad \text{as } |z|\to 1.$$

(ii)

$$\lim_{r\to 1^-} \iint_{\mathbb{D}} P(\mu_r)\bar{\nu}\,\rho(z)d^2z = \iint_{\mathbb{D}} \mu\bar{\nu}\,\rho(z)d^2z.$$

PROOF. Since

$$\iint_{\mathbb{D}} |P(\mu_r)|^2 \rho(z)d^2z = \iint_{\mathbb{D}} P(\mu_r)\bar{\mu}_r\,\rho(z)d^2z = \iint_{\mathbb{D}_r} P(\mu_r)\bar{\mu}_r\,\rho(z)d^2z < \infty,$$

$P(\mu_r) \in H^{-1,1}(\mathbb{D})$. Using $\mu(z) = -\frac{(1-|z|^2)^2}{2}\sum_{n=2}^\infty (n^3-n)a_n\bar{z}^{n-2}$ and (1.7), we get

$$P(\mu_r)(z) = -\frac{(1-|z|^2)^2}{4}\sum_{n=2}^\infty (n^3-n)a_n\left(\frac{r^{2n+2}}{n+1} - \frac{2r^{2n}}{n} + \frac{r^{2n-2}}{n-1}\right)\bar{z}^{n-2},$$

so that $(1-|z|^2)^{-2}P(\mu_r)(z)$ is continuous on $|z| = 1$.

To prove part (ii), consider the estimate

$$|(\mu - P(\mu_r))(z)| \leq \frac{3(1-|z|^2)^2}{\pi}\|\mu\|_\infty \iint_{\mathbb{D}\backslash\mathbb{D}_r} \frac{d^2u}{|1-u\bar{z}|^4}$$

$$= 3\|\mu\|_\infty (1-|z|^2)^2 \sum_{n=1}^\infty n|z|^{2n-2}(1-r^{2n})$$

$$= 3\|\mu\|_\infty \left(1 - \frac{r^2(1-|z|^2)^2}{(1-r^2|z|^2)^2}\right).$$

For fixed r the right hand side of this estimate is an increasing function of $|z|$, so that

$$\sup_{|z|\leq s} |(\mu - P(\mu_r))(z)| \leq 3\|\mu\|_\infty \left(1 - \frac{r^2(1-s^2)^2}{(1-r^2s^2)^2}\right),$$

and for fixed s,

$$\lim_{r\to 1^-} \sup_{|z|\leq s} |(\mu - P(\mu_r))(z)| = 0.$$

Also for fixed r we have the estimate
$$\|\mu - P(\mu_r)\|_\infty \leq 3\|\mu\|_\infty.$$

Now since $\nu \in L^1(\mathbb{D}, \rho(z)d^2z)$, for every $\varepsilon > 0$ there exists $0 < s < 1$ such that
$$\iint_{\mathbb{D}\setminus\mathbb{D}_s} |\nu|\rho(z)d^2z \leq \varepsilon,$$

and we obtain,
$$\left|\iint_\mathbb{D} (\mu - P(\mu_r))\,\bar\nu\rho(z)d^2z\right|$$
$$\leq \iint_{\mathbb{D}_s} |(\mu - P(\mu_r))|\,|\nu|\,\rho(z)d^2z + \iint_{\mathbb{D}\setminus\mathbb{D}_s} |(\mu - P(\mu_r))|\,|\nu|\,\rho(z)d^2z$$
$$\leq \sup_{|z|\leq s} |(\mu - P(\mu_r))(z)| \iint_\mathbb{D} |\nu|\rho(z)d^2z + 3\varepsilon\|\mu\|_\infty.$$

Passing to the limit $r \to 1^-$, we get
$$\lim_{r\to 1^-} \left|\iint_\mathbb{D} (\mu - P(\mu_r))\,\bar\nu\,\rho(z)d^2z\right| \leq 3\varepsilon\|\mu\|_\infty.$$

Since ε is arbitrary, the result follows. $\square$

LEMMA 8.2. *For $\mu, \nu \in \Omega^{-1,1}(\mathbb{D}, \Gamma)$,*
$$(\mu, \nu)_\Gamma = \lim_{r\to 1^-}\lim_{s\to 1^-} \frac{A(\Gamma\setminus\mathbb{D})}{A(\mathbb{D}_r)} \iint_\mathbb{D} P(\mu_s)\overline{P(\nu_r)}\rho(z)d^2z.$$

PROOF. Since $\mu \in \Omega^{-1,1}(\mathbb{D}, \Gamma) \subset \Omega^{-1,1}(\mathbb{D})$,
$$\iint_\mathbb{D} \mu_r \bar\nu_r \rho(z)d^2z = \iint_\mathbb{D} \mu\bar\nu_r \rho(z)d^2z = \iint_\mathbb{D} \mu\overline{P(\nu_r)}\rho(z)d^2z.$$

According to part (i) of Lemma 8.1, $P(\nu_r) \in L^1(\mathbb{D}, \rho(z)d^2z)$ for $0 < r < 1$, so that the result follows from part (ii) of Lemma 8.1. $\square$

REMARK 8.3. The limits in Lemma 8.2 can not be interchanged. Indeed, it follows from part (ii) of Lemma 8.1 that for fixed $s < 1$ the limit $r \to 1$ is always zero.

In a neighborhood of the origin in $T(\Gamma)$ the Weil-Petersson metric is given by
$$g_{\mu\bar\nu}(\kappa) = \iint_{\Gamma_\kappa\setminus\mathbb{D}} P(R(\mu, \kappa))\overline{P(R(\nu, \kappa))}\rho(z)d^2z,$$

where $\kappa \in \Omega^{-1,1}(\mathbb{D}, \Gamma)$, $\|\kappa\|_\infty$ is sufficiently small, and $\Gamma_\kappa = w_\kappa \circ \Gamma \circ w_\kappa^{-1}$.

8. FINITE-DIMENSIONAL TEICHMÜLLER SPACES

LEMMA 8.4. *Let $\mu, \nu \in \Omega^{-1,1}(\mathbb{D}, \Gamma)$. For $\kappa \in \Omega^{-1,1}(\mathbb{D}, \Gamma)$, $\|\kappa\|_\infty$ sufficiently small,*

$$g_{\mu\bar\nu}(\kappa) = \lim_{r \to 1^-} \lim_{s \to 1^-} \frac{A(\Gamma\backslash\mathbb{D})}{A(\mathbb{D}_r)} \iint_{\mathbb{D}} P(R(P(\mu_s), \kappa))\overline{P(R(P(\nu_r), \kappa))}\rho(z)d^2z.$$

PROOF. First, we have

$$\iint_{\mathbb{D}} P(R(P(\mu_s), \kappa))\overline{P(R(P(\nu_r), \kappa))}\rho(z)d^2z$$

$$= \frac{12}{\pi} \iint_{\mathbb{D}} \iint_{\mathbb{D}} \frac{P(\mu_s)(u)(w_\kappa)_u(u)^2 \overline{P(\nu_r)(z)(w_\kappa)_z(z)^2}}{(1 - w_\kappa(u)\overline{w_\kappa(z)})^4} d^2z d^2u.$$

Since $\rho P(\nu_r)$ is bounded on $\mathbb{D}$, and for $\|\kappa\|_\infty$ sufficiently small $(1/2)\rho \leq w_\kappa^* \rho \leq (3/2)\rho$, we conclude that $\rho R(P(\nu_r), \kappa)$ is also bounded on $\mathbb{D}$. As a result,

$$\iint_{\mathbb{D}} \left| \iint_{\mathbb{D}} \frac{(w_\kappa)_u(u)^2 \overline{P(\nu_r)(z)(w_\kappa)_z(z)^2}}{(1 - w_\kappa(u)\overline{w_\kappa(z)})^4} d^2z \right| d^2u$$

$$= \iint_{\mathbb{D}} \left| \iint_{\mathbb{D}} \frac{\overline{R(P(\nu_r), \kappa)(z)}}{(1 - u\bar z)^4} d^2z \right| \frac{d^2u}{1 - |\kappa(u)|^2}$$

$$\leq C \iint_{\mathbb{D}} \iint_{\mathbb{D}} \frac{(1 - |z|^2)^2}{|1 - u\bar z|^4} d^2z d^2u = \pi^2 C < \infty.$$

It follows from part (ii) of Lemma 8.1 that

$$\lim_{s \to 1^-} \iint_{\mathbb{D}} P(R(P(\mu_s), \kappa))\overline{P(R(P(\nu_r), \kappa))}\rho(z)d^2z$$

$$= \frac{12}{\pi} \iint_{\mathbb{D}} \iint_{\mathbb{D}} \frac{\mu(u)(w_\kappa)_u(u)^2 \overline{P(\nu_r)(z)(w_\kappa)_z(z)^2}}{(1 - w_\kappa(u)\overline{w_\kappa(z)})^4} d^2z d^2u$$

$$= \iint_{\mathbb{D}} P(R(\mu, \kappa))\overline{P(R(P(\nu_r), \kappa))}\rho(z)d^2z.$$

Now

$$\iint_{\mathbb{D}} P(R(\mu, \kappa))\overline{P(R(P(\nu_r), \kappa))}\rho(z)d^2z$$

$$= \frac{144}{\pi^2} \iint_{\mathbb{D}} \iint_{\mathbb{D}_r} \iint_{\mathbb{D}} \frac{\mu(u)(w_\kappa)_u(u)^2 \overline{\nu(v)(w_\kappa)_z(z)^2}}{(1 - w_\kappa(u)\overline{w_\kappa(z)})^4 (1 - z\bar v)^4} \rho(z)^{-1} d^2u d^2v d^2z$$

$$= \frac{144}{\pi^2} \iint_{\mathbb{D}_r} \lambda(v)\overline{\nu(v)}\rho(v)d^2v,$$

where

$$\lambda(v) = \rho(v)^{-1} \iint_{\mathbb{D}} \iint_{\mathbb{D}} \frac{\mu(u)(w_\kappa)_u(u)^2 \overline{(w_\kappa)_z(z)^2}\rho(z)^{-1}}{(1 - w_\kappa(u)\overline{w_\kappa(z)})^4 (1 - z\bar v)^4} d^2u d^2z.$$

Since $\mu \in \Omega^{-1,1}(\mathbb{D}, \Gamma)$ and $w_\kappa \circ \Gamma \circ w_\kappa^{-1} = \Gamma_\kappa \subseteq \mathrm{PSU}(1,1)$, it is easy to see that $\lambda \in \Omega^{-1,1}(\mathbb{D}, \Gamma)$. Using Lemma 1.9, we get

$$\lim_{r \to 1^-} \frac{A(\Gamma \backslash \mathbb{D})}{A(\mathbb{D}_r)} \iint_{\mathbb{D}} P(R(\mu, \kappa)) \overline{P(R(P(\nu_r), \kappa))} \rho(z) d^2 z$$

$$= \frac{144}{\pi^2} \iint_{\Gamma \backslash \mathbb{D}} \lambda(v) \overline{\nu(v)} \rho(v) d^2 v$$

$$= \frac{144}{\pi^2} \iint_{\mathbb{D}} \iint_{\Gamma \backslash \mathbb{D}} \iint_{\mathbb{D}} \frac{\mu(u)(w_\kappa)_u(u)^2 \overline{\nu(v)(w_\kappa)_z(z)^2}}{(1 - w_\kappa(u)\overline{w_\kappa(z)})^4 (1 - z\bar{v})^4} \rho(z)^{-1} d^2 u d^2 v d^2 z$$

$$= \frac{144}{\pi^2} \iint_{\Gamma \backslash \mathbb{D}} \iint_{\mathbb{D}} \iint_{\mathbb{D}} \frac{\mu(u)(w_\kappa)_u(u)^2 \overline{\nu(v)(w_\kappa)_z(z)^2}}{(1 - w_\kappa(u)\overline{w_\kappa(z)})^4 (1 - z\bar{v})^4} \rho(z)^{-1} d^2 u d^2 v d^2 z$$

$$= \iint_{\Gamma \backslash \mathbb{D}} P(R(\mu, \kappa)) \overline{P(R(\nu, \kappa))} \rho(z) d^2 z,$$

where we have used the fact that the integrals above do not change if we let any one of the integration variables to range over $\Gamma \backslash \mathbb{D}$ while others range over $\mathbb{D}$ (cf. [**Ahl62**]). The latter property follows from the fact that μ, ν and κ are $(-1, 1)$ tensors for Γ, and the representation $\mathbb{D} = \bigcup_{\gamma \in \Gamma} \gamma(\Gamma \backslash \mathbb{D})$. □

THEOREM 8.5. *For* $\mu, \nu, \kappa \in \Omega^{-1,1}(\mathbb{D}, \Gamma)$,

$$\left. \frac{\partial}{\partial \varepsilon} \right|_{\varepsilon=0} g_{\mu\bar{\nu}}(\varepsilon \kappa) = 0,$$

$$\left. \frac{\partial^2}{\partial \varepsilon \partial \bar{\varepsilon}} \right|_{\varepsilon=0} g_{\mu\bar{\nu}}(\varepsilon \kappa) = (\mu \bar{\kappa}, G_\Gamma(\nu \bar{\kappa}))_\Gamma + (\mu \bar{\nu}, G_\Gamma(|\kappa|^2))_\Gamma.$$

PROOF. We will use Lemma 8.4 and Theorems 7.4 and 7.7, provided one can interchange $\frac{\partial}{\partial \varepsilon}, \frac{\partial^2}{\partial \varepsilon \partial \bar{\varepsilon}}$ with the limits. This can be done as in [**Ahl62**] by showing that limits of corresponding derivatives converge uniformly on ε in a neighborhood of 0. We omit these standard arguments and concentrate on actual computations.

For the first variation of the Weil-Petersson metric we get

$$\left. \frac{\partial}{\partial \varepsilon} \right|_{\varepsilon=0} g_{\mu\bar{\nu}}(\varepsilon \kappa) = \lim_{r \to 1^-} \lim_{s \to 1^-} \frac{A(\Gamma \backslash \mathbb{D})}{A(\mathbb{D}_r)} \left. \frac{\partial}{\partial \varepsilon} \right|_{\varepsilon=0} g_{P(\mu_s)\overline{P(\nu_r)}}(\varepsilon \kappa).$$

Since $P(\mu_s), P(\nu_r) \in H^{-1,1}(\mathbb{D})$, we conclude from Theorem 7.4 that this is identically zero.

Similarly, for the second variation we have

$$\left. \frac{\partial^2}{\partial \varepsilon \partial \bar{\varepsilon}} \right|_{\varepsilon=0} g_{\mu\bar{\nu}}(\varepsilon \kappa) = \lim_{r \to 1^-} \lim_{s \to 1^-} \frac{A(\Gamma \backslash \mathbb{D})}{A(\mathbb{D}_r)} \left. \frac{\partial^2}{\partial \varepsilon \partial \bar{\varepsilon}} \right|_{\varepsilon=0} g_{P(\mu_s)\overline{P(\nu_r)}}(\varepsilon \kappa).$$

Since $P(\nu_r), P(\mu_s) \in H^{-1,1}(\mathbb{D})$ and $\kappa \in \Omega^{-1,1}(\mathbb{D}, \Gamma) \subset \Omega^{-1,1}(\mathbb{D})$, we get from Theorem 7.7[6],

$$\left.\frac{\partial^2}{\partial\varepsilon\partial\bar\varepsilon}\right|_{\varepsilon=0} g_{\mu\bar\nu}(\varepsilon\kappa)$$
$$= \lim_{r\to 1^-}\lim_{s\to 1^-}\frac{A(\Gamma\backslash\mathbb{D})}{A(\mathbb{D}_r)}\left(\left(P(\mu_s)\bar\kappa, G(P(\nu_r)\bar\kappa)\right) + \left(P(\mu_s)\overline{P(\nu_r)}, G(|\kappa|^2)\right)\right).$$

By properties **RK2** and **RK4** in Section 1.4,

$$\iint_{\mathbb{D}} |G(P(\nu_r)\kappa)|\,|\kappa|\rho(z)d^2z \leq \|\kappa\|_\infty^2 \iint_{\mathbb{D}}\iint_{\mathbb{D}} G(z,w)\,|P(\nu_r)(w)|\,\rho(z)\rho(w)d^2wd^2z$$
$$= \|\kappa\|_\infty^2 \iint_{\mathbb{D}} |P(\nu_r)(w)|\,\rho(w)d^2w < \infty,$$

and by property **RK3** $G(|\kappa|^2)$ is bounded on $\mathbb{D}$, so that it follows from Lemma 8.1 that

$$\left.\frac{\partial^2}{\partial\varepsilon\partial\bar\varepsilon}\right|_{\varepsilon=0} g_{\mu\bar\nu}(\varepsilon\kappa) = \lim_{r\to 1^-}\frac{A(\Gamma\backslash\mathbb{D})}{A(\mathbb{D}_r)}\left(\left(\mu\bar\kappa, G(P(\nu_r)\bar\kappa)\right) + \left(\mu\overline{P(\nu_r)}, G(|\kappa|^2)\right)\right).$$

We have

$$\left(\mu\bar\kappa, G(P(\nu_r)\bar\kappa)\right) = \iint_{\mathbb{D}_r} \lambda_1(v)\overline{\nu(v)}\rho(v)d^2v,$$
$$\left(\mu\overline{P(\nu_r)}, G(|\kappa|^2)\right) = \iint_{\mathbb{D}_r} \lambda_2(v)\overline{\nu(v)}\rho(v)d^2v,$$

where

$$\lambda_1(v) = \frac{12}{\pi}\rho(v)^{-1}\iint_{\mathbb{D}}\iint_{\mathbb{D}}\frac{\mu(z)\overline{\kappa(z)}G(z,u)\kappa(u)}{(1-u\bar v)^4}\rho(z)d^2ud^2z,$$
$$\lambda_2(v) = \frac{12}{\pi}\rho(v)^{-1}\iint_{\mathbb{D}}\iint_{\mathbb{D}}\frac{\mu(z)G(z,u)|\kappa(u)|^2}{(1-z\bar v)^4}\rho(u)d^2ud^2z,$$

and $\lambda_1, \lambda_2 \in \Omega^{-1,1}(\mathbb{D},\Gamma)$. It follows from Lemma 1.9 that

$$\left.\frac{\partial^2}{\partial\varepsilon\partial\bar\varepsilon}\right|_{\varepsilon=0} g_{\mu\bar\nu}(\varepsilon\kappa) = \iint_{\Gamma\backslash\mathbb{D}} \lambda_1(v)\overline{\nu(v)}\rho(v)d^2v + \iint_{\Gamma\backslash\mathbb{D}} \lambda_2(v)\overline{\nu(v)}\rho(v)d^2v$$
$$= \frac{12}{\pi}\iint_{\mathbb{D}}\iint_{\Gamma\backslash\mathbb{D}}\iint_{\mathbb{D}} \mu(z)\overline{\kappa(z)}G(z,u)\kappa(u)\frac{\overline{\nu(v)}}{(1-u\bar v)^4}\rho(z)d^2ud^2vd^2z$$
$$+ \frac{12}{\pi}\iint_{\mathbb{D}}\iint_{\Gamma\backslash\mathbb{D}}\iint_{\mathbb{D}} \mu(z)G(z,u)|\kappa(u)|^2\frac{\overline{\nu(v)}}{(1-z\bar v)^4}\rho(u)d^2ud^2vd^2z$$
$$= I_1 + I_2.$$

---

[6]It is for this case that we need the condition $\kappa \in \Omega^{-1,1}(\mathbb{D})$ in Theorem 7.7.

As before, the integrals above do not change if we let any one of the integration variables to range over $\Gamma\backslash\mathbb{D}$ while others range over $\mathbb{D}$. We have, using property **RK1**, (1.7) and (1.20),

$$I_1 = \frac{12}{\pi}\iint_{\mathbb{D}}\iint_{\mathbb{D}}\iint_{\Gamma\backslash\mathbb{D}}\mu(z)\overline{\kappa(z)}G(z,u)\kappa(u)\frac{\overline{\nu(v)}}{(1-u\bar{v})^4}\rho(z)d^2ud^2vd^2z$$

$$= \iint_{\mathbb{D}}\iint_{\Gamma\backslash\mathbb{D}}\mu(z)\overline{\kappa(z)}G(z,u)\kappa(u)\overline{\nu(u)}\rho(u)\rho(z)d^2ud^2z$$

$$= \iint_{\Gamma\backslash\mathbb{D}}\iint_{\Gamma\backslash\mathbb{D}}\mu(z)\overline{\kappa(z)}G_\Gamma(z,u)\kappa(u)\overline{\nu(u)}\rho(u)\rho(z)d^2ud^2z$$

$$= (\mu\bar{\kappa}, G_\Gamma(\nu\bar{\kappa}))_\Gamma.$$

Similarly,

$$I_2 = \iint_{\mathbb{D}}\iint_{\Gamma\backslash\mathbb{D}}\mu(z)\overline{\nu(z)}G(z,u)|\kappa(u)|^2\rho(u)\rho(z)d^2ud^2z$$

$$= \iint_{\Gamma\backslash\mathbb{D}}\iint_{\Gamma\backslash\mathbb{D}}\mu(z)\overline{\nu(z)}G_\Gamma(z,u)|\kappa(u)|^2\rho(u)\rho(z)d^2ud^2z$$

$$= (\mu\bar{\nu}, G_\Gamma(|\kappa|^2))_\Gamma,$$

and the assertion follows. $\square$

REMARK 8.6. Theorem 8.5 was proved by Wolpert [**Wol86**], and all results on Ricci, sectional, and scalar curvatures for finite-dimensional Teichmüller spaces follow from it.

We conclude this section by deriving a formula for Ricci tensor different from [**Wol86**], and indicating its application. Let $\mu_1,\ldots,\mu_d$ be an orthonormal basis of $\Omega^{-1,1}(\mathbb{D},\Gamma)$, which is a subspace of the Hilbert space $L^2(\mathbb{D},\Gamma)$ of Beltrami differentials μ for Γ such that $|\mu|\in L^2(\Gamma\backslash\mathbb{D},\rho(z)d^2z)$. Let $P:L^2(\mathbb{D},\Gamma)\to\Omega^{-1,1}(\mathbb{D},\Gamma)$ be the orthogonal projector. It follows from the definition and representation (1.7) that P is an integral operator with kernel

$$P(z,w) = \sum_{n=1}^{d}\mu_n(z)\overline{\mu_n(w)} = \frac{12}{\pi}\rho(z)^{-1}\rho(w)^{-1}\sum_{\gamma\in\Gamma}\frac{\gamma'(w)^2}{(1-\bar{z}\gamma(w))^4}.$$

8. FINITE-DIMENSIONAL TEICHMÜLLER SPACES

The Ricci tensor at the origin of $T(\Gamma)$ is given by

$$(8.1) \quad \mathcal{R}_{\mu\bar{\nu}} = \sum_{n=1}^{d} R_{\mu\bar{\mu}_n\mu_n\bar{\nu}} = -\sum_{n=1}^{d} \left((\mu\bar{\nu}, G_\Gamma(|\mu_n|^2))_\Gamma + (\mu\bar{\mu}_n, G_\Gamma(\nu\bar{\mu}_n))_\Gamma \right)$$

$$= -\sum_{n=1}^{d} \left(\iint_{\Gamma\backslash\mathbb{D}} \iint_{\Gamma\backslash\mathbb{D}} \mu(z)\overline{\nu(z)} G_\Gamma(z,w) |\mu_n(w)|^2 \rho(w)\rho(z) d^2w d^2z \right.$$

$$\left. + \iint_{\Gamma\backslash\mathbb{D}} \iint_{\Gamma\backslash\mathbb{D}} \mu(z)\overline{\mu_n(z)} G_\Gamma(z,w) \mu_n(w)\overline{\nu(w)} \rho(w)\rho(z) d^2w d^2z \right)$$

$$= -\frac{12}{\pi} \iint_{\mathbb{D}} \iint_{\Gamma\backslash\mathbb{D}} \mu(z)\overline{\nu(z)} G(z,w) \sum_{\gamma\in\Gamma} \frac{\gamma'(w)^2}{(1-\bar{w}\gamma(w))^4} \rho(w)^{-1}\rho(z) d^2w d^2z$$

$$-\frac{12}{\pi} \iint_{\Gamma\backslash\mathbb{D}} \iint_{\mathbb{D}} \mu(z)\overline{\nu(w)} G(z,w) \sum_{\gamma\in\Gamma} \frac{\gamma'(z)^2}{(1-\bar{w}\gamma(z))^4} d^2w d^2z,$$

where nothing is changed if we let any of the integration variables to range over $\Gamma\backslash\mathbb{D}$ while other range over $\mathbb{D}$.

It is instructive to compare the Ricci curvatures of the finite-dimensional Teichmüller space $T(\Gamma)$ and that of the universal Teichmüller space $T(1)$.

First, $T(\Gamma)$ is no longer a Kähler-Einstein manifold. Second, the sum over Γ in (8.1) can be transformed into a sum over the conjugacy classes of Γ. As is in [**TZ91**], using variational formulas for the Selberg zeta-function, we find that the contribution of the hyperbolic conjugacy classes is the second variation of the Selberg zeta-function at $s = 2$. The contribution of parabolic conjugacy classes (if they are present) yields a new Kähler metric on $T(\Gamma)$, introduced in [**TZ91**]. The contribution of the identity element, as it follows from Theorem 7.11, is

$$-\frac{3}{4\pi} \iint_{\mathbb{D}} \iint_{\Gamma\backslash\mathbb{D}} \mu(z)\overline{\nu(z)} G(z,w) \rho(w)\rho(z) d^2w d^2z$$

$$-\frac{12}{\pi} \iint_{\Gamma\backslash\mathbb{D}} \iint_{\mathbb{D}} \mu(z)\overline{\nu(w)} G(z,w) \frac{1}{(1-z\bar{w})^4} d^2w d^2z = -\frac{13}{12\pi}(\mu,\nu)_\Gamma.$$

As a result, we obtain a local index theorem for families of $\bar{\partial}$-operators acting on quadratic differentials on Riemann surfaces, proved in [**TZ91**]. The above arguments interpret it as an "averaged form" of Theorem 7.11. Detailed derivation of the local index theorem for families from (8.1) will be presented elsewhere.

CHAPTER 2

Kähler Potential and Period Mapping

1. Hilbert spaces and univalent functions

It is well-known (see, e.g., Section 1.2 in Chapter I) that the universal Teichmüller space $T(1)$ is isomorphic to the space $\mathcal{D}$ of univalent functions on $\mathbb{D}$. Here we characterize the univalent functions associated with the Hilbert manifold $T_0(1)$.

In addition to the Hilbert spaces $A_2(\mathbb{D})$ and $A_2(\mathbb{D}^*)$, introduced in Section 2 in Chapter 1, we define the following Hilbert spaces of holomorphic functions,

$$A_2^1(\mathbb{D}) = \left\{ \psi \text{ holomorphic on } \mathbb{D} : \|\psi\|_2^2 = \iint_{\mathbb{D}} |\psi(z)|^2 \, d^2z < \infty \right\},$$

$$A_2^1(\mathbb{D}^*) = \left\{ \psi \text{ holomorphic on } \mathbb{D}^* : \|\psi\|_2^2 = \iint_{\mathbb{D}^*} |\psi(z)|^2 \, d^2z < \infty \right\}.$$

We denote by $\overline{A_2^1(\mathbb{D})}$ and $\overline{A_2^1(\mathbb{D}^*)}$ the corresponding Hilbert spaces of antiholomorphic functions.

REMARK 1.1. Every $\psi \in A_2^1(\mathbb{D})$ corresponds to a holomorphic 1-form $\omega = \psi(z)dz$ on $\mathbb{D}$ (or on $\Gamma\backslash\mathbb{D}$ for a cofinite Fuchsian group Γ) such that the $(1,1)$-form $\omega \wedge \bar{\omega}$ is integrable. Similarly, every $\phi \in A_2(\mathbb{D})$ corresponds to a holomorphic quadratic differential $q = \phi(z)(dz)^2$ on $\mathbb{D}$ (or on $\Gamma\backslash\mathbb{D}$) such that the $(1,1)$-form $(|\phi(z)|^2/\rho(z))dz \wedge d\bar{z}$ is integrable, so that the latter space could be also denoted by $A_2^2(\mathbb{D})$. We will use the same notation $\|\ \|_2$ for the norms in these Hilbert spaces. To avoid confusion, in the main text we always denote elements in the spaces A_2 by ϕ, and elements in the spaces A_2^1 by ψ.

In addition to the Banach spaces $A_\infty(\mathbb{D})$ and $A_\infty(\mathbb{D}^*)$ introduced in Section 1.1 of Chapter 1, we define the following Banach spaces of holomorphic functions,

$$A_\infty^1(\mathbb{D}) = \left\{ \psi \text{ holomorphic on } \mathbb{D} : \|\psi\|_\infty = \sup_{z \in \mathbb{D}} \left| (1-|z|^2)\psi(z) \right| < \infty \right\},$$

$$A_\infty^1(\mathbb{D}^*) = \left\{ \psi \text{ holomorphic on } \mathbb{D}^* : \|\psi\|_\infty = \sup_{z \in \mathbb{D}^*} \left| (1-|z|^2)\psi(z) \right| < \infty \right\}.$$

For a holomorphic function $f : \Omega \to \mathbb{C}$ such that $f' \neq 0$ on Ω we set

$$\mathcal{A}(f) = \frac{f''}{f'}.$$

REMARK 1.2. Classical distortion theorem (see e.g., [**Pom92, Dur83**]) implies that if $f : \mathbb{D} \to \mathbb{C}$ and $g : \mathbb{D}^* \to \mathbb{C}$ are univalent functions, then $\mathcal{A}(f) \in A_\infty^1(\mathbb{D})$ and

$\mathcal{A}(g) \in A^1_\infty(\mathbb{D}^*)$. In [**Teo04**], it was shown that the Bers embedding of the universal Teichmüller curve $\mathcal{T}(1)$ into $A_\infty(\mathbb{D}) \oplus \mathbb{C}$ can be factorized as the composition of two holomorphic embeddings

$$\mathcal{T}(1) \to A^1_\infty(\mathbb{D}) \to A_\infty(\mathbb{D}) \oplus \mathbb{C}.$$

Here the map $\mathcal{T}(1) \to A^1_\infty(\mathbb{D})$ is given by $\gamma = g^{-1} \circ f \mapsto \mathcal{A}(f)$ and the map $A^1_\infty(\mathbb{D}) \to A_\infty(\mathbb{D}) \oplus \mathbb{C}$ is defined as

$$\psi \mapsto \left(\psi_z - \tfrac{1}{2}\psi^2, \tfrac{1}{2}\psi(0)\right).$$

Similar to Lemma 2.1 in Chapter 1, we have

LEMMA 1.3. *The vector spaces $A^1_2(\mathbb{D})$ and $A^1_2(\mathbb{D}^*)$ are subspaces of $A^1_\infty(\mathbb{D})$ and $A^1_\infty(\mathbb{D}^*)$ respectively. The natural inclusion maps $A^1_2(\mathbb{D}) \hookrightarrow A^1_\infty(\mathbb{D})$ and $A^1_2(\mathbb{D}^*) \hookrightarrow A^1_\infty(\mathbb{D}^*)$ are bounded linear mappings of Banach spaces.*

PROOF. It is sufficient to consider only the spaces of holomorphic functions on $\mathbb{D}$. For every $\psi \in A^1_2(\mathbb{D})$ let $\psi(z) = \sum_{n=1}^\infty n a_n z^{n-1}$ be the power series expansion. Then

$$\|\psi\|_2^2 = \iint_\mathbb{D} |\psi(z)|^2 d^2z = \pi \sum_{n=1}^\infty n|a_n|^2,$$

and by Cauchy-Schwarz inequality, we have

$$|\psi(z)| \le \sum_{n=1}^\infty n|a_n||z|^{n-1} \le \left(\sum_{n=1}^\infty n|a_n|^2\right)^{1/2} \left(\sum_{n=1}^\infty n|z|^{2n-2}\right)^{1/2}$$

for every $z \in \mathbb{D}$. Using

$$\sum_{n=1}^\infty n|z|^{2n-2} = \frac{1}{(1-|z|^2)^2},$$

we get

$$\|\psi\|_\infty = \sup_{z \in \mathbb{D}} |(1-|z|^2)\psi(z)| \le \frac{1}{\sqrt{\pi}} \|\psi\|_2.$$

$\square$

Similar to Remark 2.2 in Chapter 1, we get

COROLLARY 1.4. *For every $\psi \in A^1_2(\mathbb{D})$,*

$$\lim_{|z| \to 1^-} (1-|z|^2)\psi(z) = 0.$$

Similar statement holds for every $\psi \in A^1_2(\mathbb{D}^*)$.

For a holomorphic function $f : \Omega \to \mathbb{C}$ set

$$\Psi(f) = f_z - \tfrac{1}{2}f^2.$$

If $f' \ne 0$ on Ω, then $\mathcal{S}(f) = (\Psi \circ \mathcal{A})(f)$, where $\mathcal{S}(f)$ is the Schwarzian derivative of f. In [**Teo04**] it was proved that $\Psi\left(A^1_\infty(\mathbb{D})\right) \subset A_\infty(\mathbb{D})$ and $\Psi\left(A^1_\infty(\mathbb{D}^*)\right) \subset A_\infty(\mathbb{D}^*)$. Similarly, we have the following result.

LEMMA 1.5. $\Psi\left(A^1_2(\mathbb{D})\right) \subset A_2(\mathbb{D})$ *and* $\Psi\left(A^1_2(\mathbb{D}^*)\right) \subset A_2(\mathbb{D}^*)$.

PROOF. It is sufficient to consider functions on $\mathbb{D}$. For $\psi = \sum_{n=1}^{\infty} n a_n z^{n-1}$
$\in A_2^1(\mathbb{D})$ we have

$$\iint_{\mathbb{D}} |\Psi(\psi)|^2 \rho(z)^{-1} d^2z \leq 2 \iint_{\mathbb{D}} |\psi_z(z)|^2 \rho(z)^{-1} d^2z + \frac{1}{2} \iint_{\mathbb{D}} |\psi(z)|^4 \rho(z)^{-1} d^2z.$$

For the first term, a straightforward computation gives

$$\iint_{\mathbb{D}} |\psi_z(z)|^2 \rho(z)^{-1} d^2z = \frac{\pi}{2} \sum_{n=2}^{\infty} \frac{n(n-1)}{n+1} |a_n|^2 < \tfrac{1}{2} \|\psi\|_2^2 < \infty.$$

For the second term, since $\psi \in A_\infty^1(\mathbb{D})$, we have

$$\iint_{\mathbb{D}} \rho(z)^{-1} |\psi(z)|^4 d^2z \leq \tfrac{1}{4} \|\psi\|_\infty^2 \|\psi\|_2^2 < \infty.$$

$\square$

The following theorem of Becker and Pommerenke [**BP78**] characterizes univalent functions on $\mathbb{D}$ that admit a q.c. extension to a larger domain such that the complex dilation is continuous on S^1.

THEOREM 1.6. *Let $f : \mathbb{D} \to \mathbb{C}$ be a univalent function such that $f(\mathbb{D})$ is a Jordan domain. Then the following conditions are equivalent.*

(i) *f has a q.c. extension F to $\{z : |z| < R, R > 1\}$ such that the complex dilation $\mu(z) = F_{\bar{z}}/F_z$ satisfies*

$$\lim_{|z| \to 1^+} \mu(z) = 0.$$

(ii)
$$\lim_{|z| \to 1^-} (1 - |z|^2)^2 \mathcal{S}(f)(z) = 0.$$

(iii)
$$\lim_{|z| \to 1^-} (1 - |z|^2) \mathcal{A}(f)(z) = 0.$$

In [**GS92**], Gardiner and Sullivan have studied the subgroup

$$S = \text{Möb}(S^1) \backslash \text{Homeo}_s(S^1)$$

of symmetric homeomorphisms in $QS = \text{Möb}(S^1) \backslash \text{Homeo}_{qs}(S^1) \simeq T(1)$. They proved that as a Banach submanifold of $T(1)$, S is a topological group, and that univalent functions f associated to elements in S are precisely the functions satisfying condition (ii) of Theorem 1.6.

REMARK 1.7. Actually in [**GS92**] this condition is stated as follows: for every $\varepsilon > 0$ there is a compact subset K of $\mathbb{D}$ such that $|(1 - |z|^2)^2 \mathcal{S}(f)(z)| < \varepsilon$ for $z \in \mathbb{D} \setminus K$, which is clearly equivalent to (ii).

Using this remark and Remark 2.2 in Chapter 1, we get the following statement.

COROLLARY 1.8. *The group $T_0(1)$ is a subgroup of S.*

REMARK 1.9. It is known [**GS92**] that the topological group S contains the subgroup of C^1-homeomorphisms of S^1. Similarly, the topological group $T_0(1)$ contains the subgroup of C^3-homeomorphisms. Indeed, it is known (see, e.g., [**Ham02**]) that if $\gamma \in QS$ is C^3 then corresponding f and g are of C^2 class on the boundary and all their derivatives are Holder continuous with $\alpha < 1$. From here it follows that $\mathcal{S}(f) \in A_2(\mathbb{D})$.

REMARK 1.10. For $[\mu] \in T_0(1)$ it is an interesting open problem to characterize intrinsically the corresponding map $w_\mu|_{S^1}$ and the quasi-circle $f(S^1)$, as it was done for by Gardiner and Sullivan in [**GS92**] for $[\mu] \in S$.

Another important consequence of Becker-Pommerenke Theorem is the following result.

LEMMA 1.11. *Let f and g be univalent functions on $\mathbb{D}$ and $\mathbb{D}^*$ such that $\mathcal{S}(f) \in A_2(\mathbb{D})$ and $\mathcal{S}(g) \in A_2(\mathbb{D}^*)$. Then $\mathcal{A}(f) \in A_2^1(\mathbb{D})$ and $\mathcal{A}(g) \in A_2^1(\mathbb{D}^*)$.*

PROOF. It is sufficient to consider functions on $\mathbb{D}$. If $\mathcal{S}(f) \in A_2(\mathbb{D})$, then by Remark 2.2 in Chapter 1 f satisfies the condition (ii) in Theorem 1.6 and hence it satisfies the condition (iii). In particular, there exists $r' > 0$ such that

$$(1-|z|^2)|\mathcal{A}(f)(z)| \leq 1/2 \quad \text{for all} \quad r' < |z| < 1.$$

By triangle and geometric mean inequalities,

$$\begin{aligned}|\mathcal{S}(f)(z)|^2 &\geq \left(|\mathcal{A}(f)'(z)| - \tfrac{1}{2}|\mathcal{A}(f)(z)|^2\right)^2 \\ &= |\mathcal{A}(f)'(z)|^2 + \tfrac{1}{4}|\mathcal{A}(f)(z)|^4 - |\mathcal{A}(f)'(z)||\mathcal{A}(f)(z)|^2 \\ &\geq \tfrac{1}{2}\left(|\mathcal{A}(f)'(z)|^2 - |\mathcal{A}(f)(z)|^4\right),\end{aligned}$$

so that for $r' < |z| < 1$,

(1.1) $\qquad 2(1-|z|^2)^2|\mathcal{S}(f)(z)|^2 \geq (1-|z|^2)^2|\mathcal{A}(f)'(z)|^2 - \tfrac{1}{4}|\mathcal{A}(f)(z)|^2.$

Let $\mathcal{A}(f)(z) = \sum_{n=1}^{\infty} n a_n z^{n-1}$ be the power series expansion of $\mathcal{A}(f)$ and let $\mathbb{D}_r$ be the disk of radius r. We have,

$$\iint_{\mathbb{D}_r} (1-|z|^2)^2 |\mathcal{A}(f)'(z)|^2 d^2 z = \pi \sum_{n=2}^{\infty} n^2 (n-1)^2 |a_n|^2 r^{2n} \left(\frac{r^{-2}}{n-1} - \frac{2}{n} + \frac{r^2}{n+1}\right)$$

and

$$\iint_{\mathbb{D}_r} |\mathcal{A}(f)(z)|^2 d^2 z = \pi \sum_{n=1}^{\infty} n|a_n|^2 r^{2n}.$$

Using the elementary inequality

$$(n-1)^2 \left(\frac{r^{-2}}{n-1} - \frac{2}{n} + \frac{r^2}{n+1}\right) \geq \frac{1}{2n}$$

for all $n \geq 2$ and $0 < r < 1$, we get

(1.2) $\qquad \iint_{\mathbb{D}_r} (1-|z|^2)^2 |\mathcal{A}(f)'(z)|^2 d^2 z \geq \tfrac{1}{2} \iint_{\mathbb{D}_r} |\mathcal{A}(f)(z)|^2 d^2 z - \tfrac{\pi}{2} |a_1|^2 r^2.$

Integrating the inequality (1.1) over $\mathbb{D}_r \setminus \mathbb{D}_{r'}$, and using (1.2), we get for $r > r'$,

$$2\iint_{\mathbb{D}_r\setminus\mathbb{D}_{r'}} (1-|z|^2)^2|\mathcal{S}(f)(z)|^2 d^2z \geq \iint_{\mathbb{D}_r\setminus\mathbb{D}_{r'}} \left((1-|z|^2)^2|\mathcal{A}(f)'(z)|^2 - \tfrac{1}{4}|\mathcal{A}(f)(z)|^2\right) d^2z$$

$$= \iint_{\mathbb{D}_r} (1-|z|^2)^2|\mathcal{A}(f)'(z)|^2 d^2z - \tfrac{1}{4}\iint_{\mathbb{D}_r}|\mathcal{A}(f)(z)|^2 d^2z$$

$$- \iint_{\mathbb{D}_{r'}} (1-|z|^2)^2|\mathcal{A}(f)'(z)|^2 d^2z + \tfrac{1}{4}\iint_{\mathbb{D}_{r'}}|\mathcal{A}(f)(z)|^2 d^2z$$

$$\geq \tfrac{1}{4}\iint_{\mathbb{D}_r}|\mathcal{A}(f)(z)|^2 d^2z + \tfrac{1}{4}\iint_{\mathbb{D}_{r'}}|\mathcal{A}(f)(z)|^2 d^2z$$

$$- \iint_{\mathbb{D}_{r'}} (1-|z|^2)^2|\mathcal{A}(f)'(z)|^2 d^2z - \tfrac{\pi}{2}|a_1|^2 r^2.$$

Since $\mathcal{S}(f) \in A_2(\mathbb{D})$, from this inequality we conclude that there exists $C > 0$ such that

$$\iint_{\mathbb{D}_r} |\mathcal{A}(f)(z)|^2 d^2z < C$$

for all $0 < r < 1$, i.e., $\mathcal{A}(f) \in A_2^1(\mathbb{D})$. $\square$

The following statement is the main result of this section.

THEOREM 1.12. *Let $w_\mu = g_\mu^{-1} \circ f^\mu$ be the conformal welding corresponding to $[\mu] \in T(1)$. Then $[\mu] \in T_0(1)$ if and only if one of the following conditions holds.*
  (i) $\mathcal{S}(f^\mu) \in A_2(\mathbb{D})$.
  (ii) $\mathcal{A}(f^\mu) \in A_2^1(\mathbb{D})$.
  (iii) $\mathcal{S}(g_\mu) \in A_2(\mathbb{D}^*)$.
  (iv) $\mathcal{A}(g_\mu) \in A_2^1(\mathbb{D}^*)$.

PROOF. Since under the Bers embedding $\beta(T_0(1)) = \beta(T(1)) \cap A_2(\mathbb{D})$, it follows that if $w_\mu = g_\mu^{-1} \circ f^\mu$ is the conformal welding associated to $[\mu] \in T(1)$, then $\mathcal{S}(f^\mu) \in A_2(\mathbb{D})$ if and only if $[\mu] \in T_0(1)$. Let j be the antiholomorphic inversion $z \mapsto 1/\bar{z}$. Since q.c. mapping w_μ on $\hat{\mathbb{C}}$ satisfies $j \circ w_\mu \circ j = w_\mu$, we have

$$w_\mu^{-1} = j \circ w_\mu^{-1} \circ j = (j \circ (f^\mu)^{-1} \circ j) \circ (j \circ g_\mu \circ j).$$

Thus

(1.3) $$f^{\mu^{-1}} = r \circ j \circ g_\mu \circ j \quad \text{and} \quad g_{\mu^{-1}} = r \circ j \circ f^\mu \circ j,$$

where r is the dilation $z \mapsto \overline{g'_\mu(\infty)}\, z$. Since $[\mu^{-1}] \in T_0(1)$ if and only $[\mu] \in T_0(1)$, we have $\mathcal{S}(f^\mu) \in A_2(\mathbb{D})$ if and only if $\mathcal{S}(f^{\mu^{-1}}) \in A_2(\mathbb{D})$, and hence if and only if

$$\mathcal{S}(g_\mu) = \overline{\mathcal{S}(f^{\mu^{-1}}) \circ j\, j_{\bar{z}}^2} \in A_2(\mathbb{D}^*).$$

The statement of the theorem now follows from Lemmas 1.5 and 1.11. $\square$

Let $\mathcal{T}_0(1)$ be the Teichmüller curve of $T_0(1)$, i.e., the inverse image of $T_0(1)$ under the fibration $\mathcal{T}(1) \to T(1)$ of Hilbert manifolds. It was proved in Section 3 in Chapter 1 that $\mathcal{T}_0(1)$ is a topological group. Using proofs of Lemma 1.5 and Theorem 1.11, we can easily modify the proof in the Appendix of [**Teo04**] to show that $A_2^1(\mathbb{D})$ and $A_2(\mathbb{D}) \oplus \mathbb{C}$ induce the same Hilbert manifold structure on $T_0(1)$. We leave the details to Appendix A.

Results of this section justify the following

DEFINITION 1.13. The "universal Liouville action" $\mathsf{S}_1 : T_0(1) \to \mathbb{R}$ is defined by

$$\mathsf{S}_1([\mu]) = \iint_{\mathbb{D}} |\mathcal{A}(\mathsf{f}^\mu)|^2 \, d^2z + \iint_{\mathbb{D}^*} |\mathcal{A}(\mathsf{g}_\mu)|^2 \, d^2z - 4\pi \log|\mathsf{g}'_\mu(\infty)|,$$

where $w_\mu = \mathsf{g}_\mu^{-1} \circ \mathsf{f}^\mu$ is the conformal welding corresponding to $[\mu] \in T_0(1)$.

We will prove in Section 5 that the universal Liouville action is a Kähler potential of the Weil-Petersson metric on $T_0(1)$.

REMARK 1.14. When g'_μ is continuous on S^1, the last term in the definition of S_1 can be written as

$$-2 \oint_{S^1} \log|\mathsf{g}'_\mu(e^{i\theta})| d\theta.$$

When the quasicircle $\mathsf{g}_\mu(S^1) = \mathsf{f}^\mu(S^1)$ is of C^3 class, functionals of this type were studied by Schiffer and Hawley in [**SH62**]. Here we extend the definition to quasicircles for the Hilbert manifold $T_0(1)$.

2. Grunsky operators for $T_0(1)$

2.1. Grunsky coefficients and operators.

Here we prove that Grunsky operators associated to a point in $T_0(1)$ are Hilbert-Schmidt. Suppose that $f : \mathbb{D} \to \mathbb{C}$ and $g : \mathbb{D}^* \to \mathbb{C}$ are univalent functions on $\mathbb{D}$ and $\mathbb{D}^*$ such that $f(0) = 0$, $f'(0) = 1$, $g(\infty) = \infty$, and $f(\mathbb{D}) \cap g(\mathbb{D}^*) = \emptyset$. Such univalent functions are said to form a normalized disjoint pair. The generalized Grunsky coefficients $b_{n,m}$, $n, m \in \mathbb{Z}$ of a normalized disjoint pair (f, g) are defined as follows (see e.g., [**Pom92**])

$$\log \frac{g(z) - g(\zeta)}{z - \zeta} = b_{00} - \sum_{m=1}^{\infty} \sum_{n=1}^{\infty} b_{mn} z^{-m} \zeta^{-n},$$

$$\log \frac{g(z) - f(\zeta)}{bz} = -\sum_{m=1}^{\infty} \sum_{n=0}^{\infty} b_{m,-n} z^{-m} \zeta^n,$$

$$\log \frac{f(z) - f(\zeta)}{z - \zeta} = b_{00} - \sum_{m=0}^{\infty} \sum_{n=0}^{\infty} b_{-m,-n} z^m \zeta^n.$$

By definition, $b_{00} = \log b$, where $b = g'(\infty)$. Grunsky coefficients $b_{n,m}$ are symmetric in n, m when $n, m \geq 1$ or $n, m \leq 0$, so for $n \geq 0$, $m \geq 1$, we define $b_{-n,m} = b_{m,-n}$. It is also clear that coefficients $b_{n,m}$, $|n| \geq 1$ and $|m| \geq 1$ do not changed when f and g are simultaneosly post-composed with a dilation $z \mapsto rz$.

Grunsky coefficients satisfy the generalized Grunsky inequality, due to Hummel [**Hum72**] (see also [**Pom92**]).

THEOREM 2.1. Let (f, g) be a normalized disjoint pair of univalent functions. Then for every $\lambda_{-m}, \ldots, \lambda_m \in \mathbb{C}$,

$$\sum_{k=-\infty}^{\infty} |k| \left| \sum_{l=-m}^{m} b_{kl} \lambda_l \right|^2 \leq \sum_{k=-m}^{m}{}' \frac{|\lambda_k|^2}{|k|} + 2 \operatorname{Re}\left[\bar{\lambda}_0 \sum_{l=-m}^{m} b_{0l} \lambda_l \right],$$

where the prime over the sum indicates that the term $k = 0$ is omitted. The equality for all $\lambda_{-m}, \ldots, \lambda_m$ holds if and only if the set $F = \mathbb{C} \setminus \{f(\mathbb{D}) \cup g(\mathbb{D}^*)\}$ has Lebesgue measure zero.

REMARK 2.2. For $\gamma \in \mathcal{T}(1)$ let $\gamma = g^{-1} \circ f$ be the corresponding conformal welding. Since (f, g) is a normalized disjoint pair of univalent functions and the quasicircle $\mathcal{C} = f(S^1) = g(S^1)$ has Lebesgue measure zero, corresponding Grunsky coefficients b_{mn} satisfy the generalized Grunsky equality. Setting $\lambda_0 = 1$ and $\lambda_k = 0, k \neq 0$, we get

$$2 \operatorname{Re} b_{00} = \sum_{k=-\infty}^{\infty} |k| |b_{k0}|^2.$$

Since

$$\log \frac{g(z)}{z} = b_{00} - \sum_{k=1}^{\infty} b_{k0} z^{-k}, \quad \text{and} \quad \log \frac{f(z)}{z} = -\sum_{k=1}^{\infty} b_{-k,0} z^k,$$

and $\operatorname{Re} b_{00} = \log |g'(\infty)|$, we have

$$2\pi \log |g'(\infty)| = \iint_{\mathbb{D}} \left| \frac{f'(z)}{f(z)} - \frac{1}{z} \right|^2 d^2 z + \iint_{\mathbb{D}^*} \left| \frac{g'(z)}{g(z)} - \frac{1}{z} \right|^2 d^2 z.$$

According to Theorem 4.3 in Chapter 1, this gives an integral formula for the Kähler potential of the Velling-Kirillov metric on $\mathcal{T}(1)$.

Now let (f, g) be a normalized disjoint pair of univalent functions such that the corresponding set F has Lebesgue measure zero. Putting in the generalized Grunsky equality $\lambda_0 = 0$ and rescaling $\lambda_l \mapsto \sqrt{|l|} \lambda_l$, we obtain the following equality

$$\sum_{k=-\infty}^{\infty} \left| {\sum_{l=-m}^{m}}' \sqrt{|kl|} b_{kl} \lambda_l \right|^2 = {\sum_{k=-m}^{m}}' |\lambda_k|^2.$$

By polarization, we get

(2.1) $$\sideset{}{'}\sum_{k=-\infty}^{\infty} \sideset{}{'}\sum_{l=-m}^{m} \sideset{}{'}\sum_{l'=-m}^{m} \sqrt{|kl|} b_{kl} \sqrt{|kl'|} \overline{b_{kl'}} \lambda_l \bar{\eta}_{l'} = \sideset{}{'}\sum_{k=-m}^{m} \lambda_k \bar{\eta}_k,$$

where λ_k, η_k are arbitrary complex numbers. Grunsky coefficients b_{mn} give rise to semi-infinite matrices $B_l, l = 1, 2, 3, 4$, defined by

$$(B_1)_{mn} = \sqrt{mn}\, b_{-m,-n}, \quad (B_2)_{mn} = \sqrt{mn}\, b_{-m,n},$$
$$(B_3)_{mn} = \sqrt{mn}\, b_{m,-n}, \quad (B_4)_{mn} = \sqrt{mn}\, b_{mn}.$$

From generalized Grunsky equality it immediately follows that matrices B_l define bounded linear operators on the separable Hilbert space

$$\ell^2 = \left\{ x = \{x_n\}_{n=1}^{\infty} : \sum_{n=1}^{\infty} |x_n|^2 < \infty \right\}$$

which we continue to denote by $B_l, l = 1, 2, 3, 4$. Here a linear operator A on ℓ^2 associated with the matrix $\{a_{mn}\}_{m,n=1}^{\infty}$ is given by $y = Ax$, where $y_m = \sum_{n=1}^{\infty} a_{mn} x_n$.

In terms of the operators B_l, generalized Grunsky equality (2.1) is equivalent to

$$\begin{align}
(2.2) \quad & B_1 B_1^* + B_2 B_2^* = I, \qquad B_3 B_1^* + B_4 B_2^* = 0, \\
& B_1 B_3^* + B_2 B_4^* = 0, \qquad B_3 B_3^* + B_4 B_4^* = I,
\end{align}$$

where I is the identity operator on ℓ^2 and B_l^* stands for the adjoint operator to B_l. These identities immediately imply that $\|B_l\| \leq 1$, $l = 1, 2, 3, 4$.

REMARK 2.3. The operator B_4 is the Grunsky operator associated to the univalent function g. The classical Grunsky inequality (see e.g. [**Pom92**]) can be succintly stated as $I - B_4 B_4^* \geq 0$, and $I - B_4 B_4^*$ is a positive-definite operator if and only if the complement of $g(\mathbb{D}^*)$ has positive Lebesgue measure. Similarly, B_1 is the Grunsky operator associated to the univalent function f and the classical Grunsky inequality is equivalent to $I - B_1 B_1^* \geq 0$. For the pair (f^μ, g_μ) associated to a point $[\mu] \in T(1)$, the operators $I - B_1 B_1^*$ and $I - B_4 B_4^*$ are positive-definite, so that $\|B_1\|, \|B_4\| < 1$ and $\operatorname{Ker} B_2^* = \operatorname{Ker} B_3^* = \{0\}$. Moreover, it follows from symmetry property of Grunsky coefficients that also $\operatorname{Ker} B_2 = \operatorname{Ker} B_3 = \{0\}$, so that the operators $B_2, B_3 : \ell^2 \to \ell^2$ are topological isomorphisms.

The operators B_l define a bounded linear operator $\mathbf{B}$ on the Hilbert space $\ell^2 \oplus \ell^2$ by

$$\mathbf{B} = \begin{pmatrix} B_1 & B_2 \\ B_3 & B_4 \end{pmatrix}.$$

Since

$$\mathbf{B}^* = \begin{pmatrix} B_1^* & B_3^* \\ B_2^* & B_4^* \end{pmatrix},$$

the generalized Grunsky equality can be succinctly rewritten as

$$\mathbf{B}\mathbf{B}^* = \mathbf{I},$$

where $\mathbf{I} = \begin{pmatrix} I & 0 \\ 0 & I \end{pmatrix}$ is the identity operator on $\ell^2 \oplus \ell^2$. Let J be the complex-conjugation operator on ℓ^2 defined by

$$(2.3) \qquad (Jx)_n = \bar{x}_n, \quad x = \{x_n\}_{n=1}^\infty \in \ell^2.$$

Setting $\mathbf{J} = \begin{pmatrix} J & 0 \\ 0 & J \end{pmatrix}$, we can express symmetry property of Grunsky coefficients as

$$\mathbf{B}^* = \mathbf{J}\mathbf{B}\mathbf{J}.$$

Thus

$$\mathbf{B}^*\mathbf{B} = \mathbf{J}\mathbf{B}\mathbf{J}\mathbf{B} = \mathbf{J}\mathbf{B}\mathbf{B}^*\mathbf{J} = \mathbf{I},$$

so that $\mathbf{B}$ is a unitary operator on $\ell^2 \oplus \ell^2$.

The operators B_l can be also realized as linear operators from the Hilbert spaces of antiholomorphic functions to the Hilbert spaces of holomorphic functions.

Namely, the kernels

$$K_1(z,w) = \frac{1}{\pi}\left(\frac{1}{(z-w)^2} - \frac{f'(z)f'(w)}{(f(z)-f(w))^2}\right) = \frac{1}{\pi}\sum_{n,m=1}^{\infty} nm b_{-n,-m} z^{n-1} w^{m-1},$$

$$K_2(z,w) = \frac{1}{\pi}\frac{f'(z)g'(w)}{(f(z)-g(w))^2} = \frac{1}{\pi}\sum_{n,m=1}^{\infty} nm b_{-n,m} z^{n-1} w^{-m-1},$$

$$K_3(z,w) = \frac{1}{\pi}\frac{g'(z)f'(w)}{(g(z)-f(w))^2} = \frac{1}{\pi}\sum_{n,m=1}^{\infty} nm b_{n,-m} z^{-n-1} w^{m-1},$$

$$K_4(z,w) = \frac{1}{\pi}\left(\frac{1}{(z-w)^2} - \frac{g'(z)g'(w)}{(g(z)-g(w))^2}\right) = \frac{1}{\pi}\sum_{n,m=1}^{\infty} nm b_{n,m} z^{-n-1} w^{-m-1},$$

define the linear operators K_l as follows,

$$K_1 : \overline{A_2^1(\mathbb{D})} \to A_2^1(\mathbb{D}), \qquad (K_1\psi)(z) = \iint_{\mathbb{D}} K_1(z,w)\overline{\psi(w)}d^2w,$$

$$K_2 : \overline{A_2^1(\mathbb{D}^*)} \to A_2^1(\mathbb{D}), \qquad (K_2\psi)(z) = \iint_{\mathbb{D}^*} K_2(z,w)\overline{\psi(w)}d^2w,$$

$$K_3 : \overline{A_2^1(\mathbb{D})} \to A_2^1(\mathbb{D}^*), \qquad (K_3\psi)(z) = \iint_{\mathbb{D}} K_3(z,w)\overline{\psi(w)}d^2w,$$

$$K_4 : \overline{A_2^1(\mathbb{D}^*)} \to A_2^1(\mathbb{D}^*), \qquad (K_4\psi)(z) = \iint_{\mathbb{D}^*} K_4(z,w)\overline{\psi(w)}d^2w.$$

REMARK 2.4. It is well-known that if ϕ is a holomorphic function on $\mathbb{D}$, then

$$\iint_{\mathbb{D}} \frac{\overline{\phi(w)}}{(z-w)^2} d^2w = 0,$$

where the integral is understood in the principal value sense. Hence we can also represent operators K_1 and K_4 by the singular kernels

$$-\frac{1}{\pi}\frac{f'(z)f'(w)}{(f(z)-f(w))^2} \quad \text{and} \quad -\frac{1}{\pi}\frac{g'(z)g'(w)}{(g(z)-g(w))^2}.$$

The Hilbert spaces $A_2^1(\mathbb{D})$ and $A_2^1(\mathbb{D}^*)$ have standard orthonormal bases $\{e_n\}_{n=1}^{\infty}$ and $\{f_n\}_{n=1}^{\infty}$, given respectively by

$$e_n(z) = \sqrt{\frac{n}{\pi}} z^{n-1} \quad \text{and} \quad f_n(z) = \sqrt{\frac{n}{\pi}} z^{-n-1}, \quad n \in \mathbb{N}.$$

These bases define isomorphisms $A_2^1(\mathbb{D}) \simeq \ell^2$ and $A_2^1(\mathbb{D}^*) \simeq \ell^2$. The operators K_l and their adjoints K_l^* — integral operators with the kernels $K_l^*(z,w) = \overline{K_l(w,z)}$, correspond respectively to the operators B_l and B_l^*, $l = 1, 2, 3, 4$. Similarly, positive self-adjoint operators $\mathsf{K}_l = K_l K_l^*$ are integral operators which correspond to the operators $B_l B_l^*$, and we denote the kernels of the operators K_l by $\mathsf{K}_l(z,w)$. Due to the relations (2.2),

(2.4) $$\mathsf{K}_2 = I - \mathsf{K}_1, \quad \mathsf{K}_3 = I - \mathsf{K}_4.$$

LEMMA 2.5. *The kernel $K_1(z,w)$ of the operator $\mathsf{K}_1 : \overline{A_2^1(\mathbb{D})} \to A_2^1(\mathbb{D})$ satisfies*

$$(2.5) \qquad \iint_\mathbb{D} \iint_\mathbb{D} |K_1(z,w)|^2 d^2z d^2w < \infty$$

if and only if the operator K_1 is Hilbert-Schmidt, i.e., if and only if the operator $\mathsf{K}_1 = K_1 K_1^*$ on $A_2^1(\mathbb{D})$ is of trace class. In this case,

$$\operatorname{Tr} \mathsf{K}_1 = \iint_\mathbb{D} \iint_\mathbb{D} |K_1(z,w)|^2 d^2z d^2w = \iint_\mathbb{D} \mathsf{K}_1(z,z) d^2z,$$

and $\mathcal{S}(f) \in A_2(\mathbb{D})$, where f is the univalent function associated with the kernel $K_1(z,w)$. Similar statements hold for the operators K_4 and K_4.

PROOF. It is sufficient to prove the lemma for the operator K_1. For the basis $\{e_n\}_{n \in \mathbb{N}}$ of the Hilbert space $A_2^1(\mathbb{D})$ we have

$$\operatorname{Tr} \mathsf{K}_1 = \sum_{n=1}^\infty \langle \mathsf{K}_1 e_n, e_n \rangle = \sum_{n=1}^\infty \|K_1^* e_n\|^2 = \sum_{n,m=1}^\infty nm |b_{-n,-m}|^2$$
$$= \iint_\mathbb{D} \iint_\mathbb{D} |K_1(z,w)|^2 d^2z d^2w = \iint_\mathbb{D} \mathsf{K}_1(z,z) d^2z.$$

Since the operator K_1 is positive, it is of trace class if and only if the inequality (2.5) holds. On the other hand, we have

$$\mathcal{S}(f)(z) = -6\pi \lim_{w \to z} K_1(z,w) = -6 \sum_{n=2}^\infty \left(\sum_{k+l=n} kl b_{-k,-l} \right) z^{n-2}.$$

Hence if the inequality (2.5) holds,

$$\|\mathcal{S}(f)\|_2^2 = 18\pi \sum_{n=2}^\infty \frac{1}{n^3 - n} \left| \sum_{k=1}^{n-1} k(n-k) b_{-k,-(n-k)} \right|^2$$
$$\leq 18\pi \sum_{n=2}^\infty \frac{1}{n^3-n} \left(\sum_{k=1}^{n-1} k(n-k) \right) \left(\sum_{k=1}^{n-1} k(n-k) |b_{-k,-(n-k)}|^2 \right)$$
$$= 3\pi \sum_{n=2}^\infty \sum_{k=1}^{n-1} k(n-k) |b_{-k,-(n-k)}|^2 = 3\pi \sum_{n,m=1}^\infty nm |b_{-n,-m}|^2 < \infty.$$

$\square$

THEOREM 2.6. *If the pair $(\mathrm{f}^\mu, \mathrm{g}_\mu)$ corresponds to a point $[\mu] \in T_0(1)$, then the operators K_1 and K_4 associated to f^μ and g_μ respectively, are of trace class.*

PROOF. According to Lemma 2.5, it is sufficient to show that

$$\iint_\mathbb{D} \mathsf{K}_1(z,z) d^2z < \infty \quad \text{and} \quad \iint_{\mathbb{D}^*} \mathsf{K}_4(z,z) d^2z < \infty.$$

For $[\mu] \in T_0(1)$ choose a representative $\mu \in L^2(\mathbb{D}^*, \rho(z) d^2z) \cap \mathcal{O}(\mathbb{D}^*)_1$. It follows from Lemma 2.9 in Chapter 1 that the path $[t\mu]$ connecting 0 to $[\mu]$ in $T(1)$ lies on $T_0(1)$. Let $w_{t\mu} = \mathrm{g}_{t\mu}^{-1} \circ \mathrm{f}^{t\mu}$ be the corresponding conformal welding and denote by $(K_1)_t(z,w)$ the kernel $K_1(z,w)$ associated with the univalent function $\mathrm{f}^{t\mu}$. We have the following lemma.

LEMMA 2.7.

$$\left.\frac{d}{ds}\right|_{s=0} (K_1)_{s+t}\left(\mathrm{f}_t^{-1}(z), \mathrm{f}_t^{-1}(w)\right) \left(\mathrm{f}_t^{-1}\right)'(z) \left(\mathrm{f}_t^{-1}\right)'(w)$$

(2.6)
$$= \frac{1}{\pi^2} \iint_{\Omega_t^*} \frac{\mu_t(u)}{(u-z)^2(u-w)^2} d^2u,$$

where $\Omega_t^* = \mathrm{f}^{t\mu}(\mathbb{D}^*) = \mathrm{g}_{t\mu}(\mathbb{D}^*)$,

$$(\mu_t \circ \mathrm{g}_{t\mu}) \frac{\overline{\mathrm{g}'_{t\mu}}}{\mathrm{g}'_{t\mu}} = D_{t\mu} R_{(t\mu)^{-1}}(\mu),$$

and the integral (2.6) is understood in the principal value sense.

PROOF. Set $w_t = w_{t\mu}, \mathrm{f}_t = \mathrm{f}^{t\mu}, \mathrm{g}_t = \mathrm{g}_{t\mu}$ and $v_s = \mathrm{f}_{s+t} \circ \mathrm{f}_t^{-1}$. We have

$$v_s \circ \mathrm{g}_t = \mathrm{g}_{s+t} \circ w_{s+t} \circ w_t^{-1},$$

so that v_s is a q.c. mapping which is holomorphic on $\Omega_t = \mathrm{f}_t(\mathbb{D})$ and has Beltrami differential $\mu_{s,t}$ on Ω_t^* with

$$(\mu_{s,t} \circ \mathrm{g}_t) \frac{\overline{\mathrm{g}'_t}}{\mathrm{g}'_t} = \frac{(w_{s+t} \circ w_t^{-1})_{\bar{z}}}{\overline{(w_{s+t} \circ w_t^{-1})_z}}.$$

It follows from the standard variational formula for q.c. mappings that

(2.7)
$$\left.\frac{d}{ds}\right|_{s=0} v_s(z) = -\frac{1}{\pi} \iint_{\Omega_t^*} \frac{\mu_t(u) z(z-1)}{(u-z)u(u-1)} d^2u + p(z),$$

where $p(z)$ is a degree two polynomial. We have

$$(K_1)_{s+t}\left(\mathrm{f}_t^{-1}(z), \mathrm{f}_t^{-1}(w)\right) \left(\mathrm{f}_t^{-1}\right)'(z) \left(\mathrm{f}_t^{-1}\right)'(w)$$
$$= \frac{1}{\pi} \frac{\left(\mathrm{f}_t^{-1}\right)'(z) \left(\mathrm{f}_t^{-1}\right)'(w)}{\left(\mathrm{f}_t^{-1}(z) - \mathrm{f}_t^{-1}(w)\right)^2} - \frac{1}{\pi} \frac{v'_s(z) v'_s(w)}{(v_s(z) - v_s(w))^2},$$

and

$$\left.\frac{d}{ds}\right|_{s=0} \frac{v'_s(z) v'_s(w)}{(v_s(z) - v_s(w))^2} = -\frac{1}{\pi} \iint_{\Omega^*} \frac{\mu_t(u)}{(u-z)^2(u-w)^2} d^2u,$$

so that the result follows. □

Now we use the fundamental theorem of calculus to estimate

$$\iint_{\mathbb{D}} \mathsf{K}_1(z,z) d^2 z = \iint_{\mathbb{D}} \iint_{\mathbb{D}} |(K_1)_1(z,w)|^2 d^2 z d^2 w$$

$$= \iint_{\mathbb{D}} \iint_{\mathbb{D}} \left| \int_0^1 \frac{d}{dt} (K_1)_t(z,w) dt \right|^2 d^2 z d^2 w$$

$$\leq \int_0^1 \iint_{\mathbb{D}} \iint_{\mathbb{D}} \left| \frac{d}{dt} (K_1)_t(z,w) \right|^2 d^2 z d^2 w dt$$

$$= \int_0^1 \iint_{\mathbb{D}} \iint_{\mathbb{D}} \left| \frac{d}{ds} \right|_{s=0} (K_1)_{t+s}(z,w) \right|^2 d^2 z d^2 w dt$$

$$= \int_0^1 I(t) dt.$$

Making a change of variables $z \mapsto \mathsf{f}_t^{-1}(z)$, $w \mapsto \mathsf{f}_t^{-1}(w)$ in the inner integral $I(t)$, we get

$$I(t) = \iint_{\Omega_t} \iint_{\Omega_t} \left| \frac{d}{ds} \right|_{s=0} (K_1)_{t+s} \left(\mathsf{f}_t^{-1}(z), \mathsf{f}_t^{-1}(w) \right) \left(\mathsf{f}_t^{-1} \right)'(z) \left(\mathsf{f}_t^{-1} \right)'(w) \right|^2 d^2 z d^2 w$$

$$= \frac{1}{\pi^4} \iint_{\Omega_t} \iint_{\Omega_t} \left| \iint_{\Omega_t^*} \frac{\mu_t(u)}{(u-z)^2(u-w)^2} d^2 u \right|^2 d^2 z d^2 w.$$

Using the inequality

(2.8) $$\iint_{\Omega_t} \frac{d^2 w}{|w-z|^4} \leq 4\pi (\rho_2)_t(z), \; z \in \Omega_t^*,$$

where $(\rho_2)_t(z)$ is the density of the hyperbolic metric on Ω_t^* (see the proof of Theorem 2.3 in Chapter 1), and the fact that the Hilbert transform is an isometry on $L^2(\mathbb{C}, d^2 z)$, we obtain

$$I(t) \leq \frac{1}{\pi^2} \iint_{\Omega_t} \iint_{\Omega_t^*} \frac{|\mu_t(z)|^2}{|z-w|^4} d^2 z d^2 w \leq \frac{4}{\pi} \iint_{\Omega_t^*} |\mu_t(z)|^2 (\rho_2)_t(z) d^2 z$$

$$= \frac{4}{\pi} \iint_{\mathbb{D}^*} |\tilde{\mu}_t(z)|^2 \rho(z) d^2 z = \frac{4}{\pi} \|\tilde{\mu}_t\|_2^2,$$

where $\tilde{\mu}_t = D_{t\mu} R_{(t\mu)^{-1}}(\mu)$. Now it follows from Remark 2.8 in Chapter 1 that there exists a constant C such that

$$\|\tilde{\mu}_t\|_2 \leq C \|\mu\|_2$$

for all $0 \leq t \leq 1$, so that

$$\iint_{\mathbb{D}} \mathsf{K}_1(z,z) d^2 z < \infty.$$

The corresponding estimate for the kernel $\mathsf{K}_4(z,w)$ is proved similarly. Alternatively, using the relation (1.3) we get

$$\mathsf{K}_1([\mu^{-1}])(z,w) = \overline{\mathsf{K}_4([\mu])\left(\frac{1}{\bar{z}},\frac{1}{\bar{w}}\right)}\frac{1}{z^2}\frac{1}{w^2}. \tag{2.9}$$

Since $T_0(1)$ is a group, the inequality for the kernel K_4 follows from the corresponding inequality for the kernel K_1. $\square$

REMARK 2.8. Actually using the generalized Grunsky equality one can prove an estimate sharper than (2.8). Just observe that for $z \in \Omega_t^*$

$$\iint\limits_{\Omega_t} \frac{d^2w}{|z-w|^4} = \pi^2 \mathsf{K}_3(g^{-1}(z), g^{-1}(z))|(g^{-1})'(z)|^2$$

and that $\frac{1}{\pi(1-z\bar{w})^2}$ is the kernel of the identity operator on $A_2^1(\mathbb{D}^*)$. Hence the second equation in (2.4) gives,

$$\mathsf{K}_3(z,z) = \frac{1}{\pi(1-|z|^2)^2} - \mathsf{K}_4(z,z) \leq \frac{1}{\pi(1-|z|^2)^2},$$

and we get

$$\iint\limits_{\Omega_t} \frac{d^2w}{|z-w|^4} \leq \frac{\pi}{4}(\rho_2)_t(z).$$

COROLLARY 2.9. *Grunsky operators B_1 and B_4 associated with the pair $(\mathrm{f}^\mu, \mathrm{g}_\mu)$, $[\mu] \in T(1)$, are Hilbert-Schmidt operators on ℓ^2 if and only if $[\mu] \in T_0(1)$.*

PROOF. Under the isomorphisms $A_2^1(\mathbb{D}) \simeq \ell^2$ and $A_2^1(\mathbb{D}^*) \simeq \ell^2$, the operators K_1 and K_4 correspond to the operators $B_1 B_1^*$ and $B_4 B_4^*$ respectively. Since $\beta(T_0(1)) = A_2(\mathbb{D}) \cap \beta(T(1))$, the "only if" part of the statement follows from Lemma 2.5. $\square$

As an application, consider the Hilbert space $\mathscr{S}_2$ of Hilbert-Schmidt operators on ℓ^2,

$$\mathscr{S}_2 = \left\{ T : \ell^2 \to \ell^2 \text{ a bounded operator} \ \Big| \ \|T\|_2^2 = \operatorname{Tr} TT^* < \infty \right\},$$

and define the mapping $\mathscr{P} : T_0(1) \to \mathscr{S}_2$ by

$$\mathscr{P}([\mu]) = B_1(\mathrm{f}^\mu), \ [\mu] \in T_0(1).$$

Since Grunsky coefficients characterize univalent functions up to a post-composition with Möbius transformation, the mapping $\mathscr{P}$ is one to one. In fact, we have a stronger result.

THEOREM 2.10. *The mapping $\mathscr{P}$ is a holomorphic inclusion of the Hilbert manifold $T_0(1)$ into the Hilbert space $\mathscr{S}_2$.*

PROOF. We need to show that for every $[\nu] \in T_0(1)$ and $\mu \in H^{-1,1}(\mathbb{D}^*)$, the map $\mathbb{C} \ni t \mapsto B_1(t) = B_1(f^{\nu+t\mu})$ is holomorphic in a neighbourhood of $t=0$ in $\mathbb{C}$. For this aim, since the mapping $[\mu] \to \mathrm{f}^\mu(z)$ is holomorphic for fixed $z \in \mathbb{D}$, for every $z, w \in \mathbb{D}$ the map

$$t \mapsto K_1^{\nu+t\mu}(z,w) = \frac{1}{\pi}\left(\frac{1}{(z-w)^2} - \frac{(\mathrm{f}^{\nu+t\mu})'(z)(\mathrm{f}^{\nu+t\mu})'(w)}{(\mathrm{f}^{\nu+t\mu}(z) - \mathrm{f}^{\nu+t\mu}(w))^2}\right)$$

is holomorphic in a neighbourhood of $t = 0$ in $\mathbb{C}$. We choose $\delta > 0$ so that $\|\nu + t\mu\|_\infty < 1$ for all $|t| < \delta$. For every t_0 such that $|t_0| < \delta$, let δ_1 be such that $0 < \delta_1 < \delta - |t_0|$. Then for all $|t - t_0| < \delta_1$, we have by Cauchy integral formula,

$$\left(K_1^{\nu+t\mu} - K_1^{\nu+t_0\mu} - (t - t_0) \left.\frac{d}{dt}\right|_{t=t_0} K_1^{\nu+t\mu} \right)(z, w)$$
$$= \frac{(t - t_0)^2}{2\pi i} \oint_{|\zeta - t_0| = \delta_1} \frac{K_1^{\nu+\zeta\mu}(z, w)}{(\zeta - t)(\zeta - t_0)^2} d\zeta.$$

Hence
(2.10)
$$\left\| \frac{B_1(\mathbf{f}^{\nu+t\mu}) - B_1(\mathbf{f}^{\nu+t_0\mu})}{t - t_0} - \left.\frac{d}{dt}\right|_{t=t_0} B_1(\mathbf{f}^{\nu+t\mu}) \right\|_2^2$$
$$= \iint_{\mathbb{D}} \iint_{\mathbb{D}} \left| \left(\frac{K_1^{\nu+t\mu} - K_1^{\nu+t_0\mu}}{t - t_0} - \left.\frac{d}{dt}\right|_{t=t_0} K_1^{\nu+t\mu} \right)(z, w) \right|^2 d^2z\, d^2w$$
$$\leq \frac{|t - t_0|^2}{4\pi^2} \oint_{|\zeta - t_0| = \delta_1} \iint_{\mathbb{D}} \iint_{\mathbb{D}} \left| K_1^{\nu+\zeta\mu}(z, w) \right|^2 d^2z\, d^2w\, |d\zeta| \oint_{|\zeta - t_0| = \delta_1} \frac{|d\zeta|}{|\zeta - t|^2 |\zeta - t_0|^4}.$$

We have from the proof of Theorem 2.6,
$$\iint_{\mathbb{D}} \iint_{\mathbb{D}} \left| K_1^{\nu+\zeta\mu}(z, w) \right|^2 d^2z\, d^2w \leq C \|\nu + \zeta\mu\|_2^2 \leq C(\|\nu\|_2 + \delta_1 \|\mu\|_2)^2,$$

so that (2.10) tends to 0 as $t \to t_0$, which proves the assertion. □

REMARK 2.11. Since the classical Grunsky operator B_1 is bounded, the mapping $\mathscr{P}$ extends to the whole Banach manifold $T(1)$. Let $\mathscr{B}(\ell^2)$ be the space of bounded linear operators on ℓ^2,

$$\mathscr{B}(\ell^2) = \left\{ T : \ell^2 \to \ell^2 \text{ a linear operator} : \|T\| = \sup_{\|u\|=1} \|Tu\| < \infty. \right\},$$

and define the mapping $\hat{\mathscr{P}} : T(1) \to \mathscr{B}(\ell^2)$ by

$$\hat{\mathscr{P}}([\mu]) = B_1(\mathbf{f}^\mu), \quad [\mu] \in T(1).$$

Analogous to Theorem 2.10, we show in Appendix B that the mapping $\hat{\mathscr{P}}$ is a holomorphic inclusion.

2.2. Fredholm eigenvalues and Fredholm determinant. In his paper [**Sch57**], Schiffer has studied the eigenvalues of the classical Poincaré-Fredholm boundary value problem of potential theory on a C^3 curve. Here we show how Fredholm eigenvalues for a quasi-circle $\mathcal{C} = \mathbf{f}^\mu(S^1) = \mathbf{g}_\mu(S^1)$, associated with $[\mu] \in T_0(1)$, are related to the eigenvalues of trace class operators K_1 and K_4.

Let $\mathfrak{h}$ be a separable Hilbert space with the inner product $\langle\,,\,\rangle$. A conjugation operator J on $\mathfrak{h}$ is an $\mathbb{R}$-linear operator satisfying $J^2 = I$ and

$$(Jx, Jy) = \overline{(x, y)} \quad \text{for all} \quad x, y \in \mathfrak{h}.$$

Conjugation operator is necessarily complex anti-linear. For every bounded linear operator T on $\mathfrak{h}$,
$$\langle JTJx, y\rangle = \overline{\langle TJx, Jy\rangle} = \overline{\langle Jx, T^*Jy\rangle} = \langle x, JT^*Jy\rangle \quad \text{for all} \quad x, y \in \mathfrak{h},$$
so that
$$(JTJ)^* = JT^*J.$$
In particular, if U is a unitary operator on $\mathfrak{h}$, then JUJ is also a unitary operator. For a bounded linear operator T on $\mathfrak{h}$ its transpose is defined as
$$T^t = JT^*J.$$
Generalizing the notion of symmetric complex-valued matrix, a bounded operator T on $\mathfrak{h}$ is called symmetric with respect to the conjugation J, if
$$T = T^t.$$
The Hilbert space $\mathfrak{h} = \ell^2$ carries a standard conjugation operator J, defined by (2.3). The following statement is a generalization of Schur's Lemma (see, e.g., [**Pom92**, Sect. 3.6]) to the case of compact operators on ℓ^2.

LEMMA 2.12. *Let T be a compact operator on ℓ^2, symmetric with respect to the standard conjugation operator J. Then there exist a unitary operator U on ℓ^2 and an operator $D \geq 0$ on ℓ^2, diagonal with respect to the standard basis for ℓ^2, such that*
$$T = UDU^t.$$

PROOF. As in [**Pom92**], consider the decomposition
$$T = \frac{T + JTJ}{2} + i\frac{T - JTJ}{2i} = A + iB,$$
where A and B are self-adjoint compact operators satisfying $AJ = JA$ and $BJ = JB$. Let $\mathbf{T}$ be the self-adjoint operator on the Hilbert space $\ell^2 \oplus \ell^2$ defined by
$$\mathbf{T} = \begin{pmatrix} A & B \\ B & -A \end{pmatrix}.$$
The operator $\mathbf{T}$ is compact and satisfies

(2.11) $$\mathbf{TE} = -\mathbf{ET} \quad \text{and} \quad \mathbf{TJ} = \mathbf{JT},$$

where
$$\mathbf{E} = \begin{pmatrix} 0 & -I \\ I & 0 \end{pmatrix} \quad \text{and} \quad \mathbf{J} = \begin{pmatrix} J & 0 \\ 0 & J \end{pmatrix}.$$
From the first equation in (2.11) it follows that if $\mathbf{u} \in \ell^2 \oplus \ell^2$ is an eigenvector for $\mathbf{T}$ with eigenvalue λ, then $\mathbf{v} = \mathbf{Eu}$ is also an eigenvector for $\mathbf{T}$ with eigenvalue $-\lambda$. It follows from Hilbert-Schmidt theorem on canonical form of compact self-adjoint operator that there exist a unitary operator $\mathbf{U}$ on $\ell^2 \oplus \ell^2$ of the form
$$\mathbf{U} = \begin{pmatrix} U_1 & U_2 \\ U_2 & -U_1 \end{pmatrix},$$
and an operator $\mathbf{D}$ on $\ell^2 \oplus \ell^2$ of the form
$$\mathbf{D} = \begin{pmatrix} D & 0 \\ 0 & -D \end{pmatrix},$$

where D is diagonal with non-negative entries, such that
$$\mathbf{T} = \mathbf{UDU}^*.$$
From the second equation in (2.11) it follows that $\mathbf{T}(\mathbf{JUJ}) = (\mathbf{JUJ})\mathbf{D}$. Since $\mathbf{JUJ}$ is also a unitary operator, we have
$$\mathbf{T} = (\mathbf{JUJ})\mathbf{D}(\mathbf{JUJ})^*.$$
Consequently, we can choose $\mathbf{U}$ so that $\mathbf{U} = \mathbf{JUJ}$. Now it follows from the canonical form that
$$T = A + iB = (U_1 + iU_2)D(U_1^* + iU_2^*).$$
Let $U = U_1 + iU_2 : \ell^2 \to \ell^2$. Since $\mathbf{U}$ is a unitary operator, U is also unitary, and the property $\mathbf{U}^* = \mathbf{JU}^*\mathbf{J}$ implies that
$$JU^*J = J(U_1^* - iU_2^*)J = U_1^* + iU_2^*,$$
since J is complex anti-linear. □

COROLLARY 2.13. *The non-zero entries of the operator D are singular values of the operator T.*

PROOF. Since the operator U^t is unitary,
$$TT^* = UD^2U^* = UD^2U^{-1},$$
so that the entries of D^2 are the eigenvalues of TT^*. □

Now let (f, g) be a normalized disjoint pair of univalent functions such that the corresponding set F has Lebesgue measure zero and the Grunsky operator B_1 is compact. We apply Schur's Lemma to the operator B_1 on ℓ^2. It follows from the symmetry property of Grunsky coefficients that
$$B_1^* = JB_1J.$$
Thus there exist a unitary operator U on ℓ^2 and a diagonal operator D with non-negative entries such that
$$B_1 = UDU^t.$$
From the first identity in (2.2), we obtain
$$U^{-1}B_2B_2^*U = I - D^2.$$
Since $\|B_1\| < 1$, the operator $I - D^2$ is positive-definite and hence invertible, so that the operator
$$V = B_2^*U(I - D^2)^{-1/2}$$
is also unitary. Using the property $B_3^t = B_2$, which follows from the symmetry of Grunsky coefficients, and the third identity in (2.2), we obtain
$$V^tB_4V = -D.$$
Collecting everything together, we get the following identities:
$$B_1JUJ = UD, \qquad B_3JUJ = JVJ(I - D^2)^{1/2},$$
$$B_2V = U(I - D^2)^{1/2}, \qquad B_4V = -JVJD.$$

2. GRUNSKY OPERATORS FOR $T_0(1)$

Letting

$$\lambda_n = (D)_{nn}, \qquad \rho_n = ((1-D^2)^{1/2})_{nn} = \sqrt{1-\lambda_n^2}$$

$$\mathfrak{u}_n(z) = \sum_{m=1}^{\infty} \sqrt{\frac{m}{\pi}} U_{mn} z^{m-1}, \qquad \mathfrak{v}_n(z) = \sum_{m=1}^{\infty} \sqrt{\frac{m}{\pi}} (JVJ)_{mn} z^{-m-1},$$

and realizing B_l's as linear operators K_l's, we obtain for $n \in \mathbb{N}$,

$$\iint_{\mathbb{D}} K_1(z,w)\overline{\mathfrak{u}_n(w)} d^2w = \lambda_n \mathfrak{u}_n(z), \qquad \iint_{\mathbb{D}} K_3(z,w)\overline{\mathfrak{u}_n(w)} d^2w = \rho_n \mathfrak{v}_n(z)$$

$$\iint_{\mathbb{D}^*} K_2(z,w)\overline{\mathfrak{v}_n(w)} d^2w = \rho_n \mathfrak{u}_n(z), \qquad \iint_{\mathbb{D}^*} K_4(z,w)\overline{\mathfrak{v}_n(w)} d^2w = -\lambda_n \mathfrak{v}_n(z).$$

Setting

$$u_n = \mathfrak{u}_n \circ f^{-1}(f^{-1})' \qquad \text{and} \qquad v_n = \mathfrak{v}_n \circ g^{-1}(g^{-1})',$$

we get,

(2.12)
$$\frac{1}{\pi} \iint_{\Omega} \frac{\overline{u_n(w)}}{(z-w)^2} d^2w = -\lambda_n u_n(z), \quad z \in \Omega,$$

$$\frac{1}{\pi} \iint_{\Omega} \frac{\overline{u_n(w)}}{(z-w)^2} d^2w = \rho_n v_n(z), \quad z \in \Omega^*,$$

$$\frac{1}{\pi} \iint_{\Omega^*} \frac{\overline{v_n(w)}}{(z-w)^2} d^2w = \rho_n u_n(z), \quad z \in \Omega,$$

$$\frac{1}{\pi} \iint_{\Omega^*} \frac{\overline{v_n(w)}}{(z-w)^2} d^2w = \lambda_n v_n(z), \quad z \in \Omega^*.$$

Comparing equations (2.12) with corresponding formulas in [**Sch57**], we find that $\{\pm \lambda_n^{-1}\}_{n=1}^{\infty}$ are Fredholm eigenvalues associated to the quasi-circle $\mathcal{C} = f(S^1) = g(S^1)$.

REMARK 2.14. The relation between the Fredholm eigenvalues and the eigenvalues of the Grunsky operator for a C^3 curve was first obtained by Schiffer in [**Sch81**]. Specifically, in [**Sch81**] Schiffer has shown that Fredholm eigenvalues, defined as the eigenvalues of classical Poincaré-Fredholm integral operator on C^3 curve, satisfy (2.12). Furthermore, using completeness of the bases $\{u_n\}$, $\{v_n\}$ in corresponding Hilbert spaces, he proved the relation (2.2), which is equivalent to the generalized Grunsky equality with $\lambda_0 = 0$. Here we use the opposite approach. We start from the generalized Grunsky equality for the pair (f^μ, g_μ) for $[\mu] \in T_0(1)$ and use it for deriving all necessary properties of the Grunsky operators. In particular, we prove that the Grunsky operators B_1 and B_4 associated with $[\mu] \in T_0(1)$ are Hilbert-Schmidt. The case we consider is more general than in [**Sch81**] since the set of all quasi-circles $f^\mu(S^1)$ for $[\mu] \in T_0(1)$ contains the set of all C^3 curves as a proper subset. In fact, we prove in Appendix B that the Grunsky operators B_1 and B_4 associated with $[\mu] \in T(1)$ are compact if and only if $[\mu] \in S$, the subgroup of symmetric homeomorphisms of S^1. Our analysis of the relation between singular values of Grunsky operators and Fredholm eigenvalues still holds for this case.

As in [**Sch59**], for a pair (f,g) such that the corresponding operators K_1 and K_4 are of trace class, we define the Fredholm determinant for the corresponding quasi-circle $\mathcal{C} = f(S^1)$ by

$$\operatorname{Det}_F(\mathcal{C}) = \prod_{n=1}^{\infty} \rho_n^2 = \det(I - \mathsf{K}_1) = \det(I - \mathsf{K}_4).$$

Theorem 2.6 justifies the following definition.

DEFINITION 2.15. The real-valued function $\mathsf{S}_2 : T_0(1) \to \mathbb{R}$ is defined as

$$\mathsf{S}_2([\mu]) = \log \operatorname{Det}_F(\mathsf{f}^\mu(S^1)), \quad [\mu] \in T_0(1).$$

It follows from (2.9) that

(2.13) $$\mathsf{S}_2([\mu]) = \mathsf{S}_2([\mu]^{-1}), \quad [\mu] \in T_0(1).$$

2.3. Period matrix of 1-forms. For a normalized disjoint pair (f,g) of univalent functions we set $\Omega = f(\mathbb{D})$, $\Omega^* = g(\mathbb{D}^*)$, and define the Hilbert spaces

$$A_2^1(\Omega) = \left\{ \psi \text{ holomorphic on } \Omega : \|\psi\|_2^2 = \iint_\Omega |\psi(z)|^2 \, d^2z < \infty \right\},$$

$$A_2^1(\Omega^*) = \left\{ \psi \text{ holomorphic on } \Omega^* : \|\psi\|_2^2 = \iint_{\Omega^*} |\psi(z)|^2 \, d^2z < \infty \right\}.$$

The Hilbert spaces $A_2^1(\Omega)$ and $A_2^1(\Omega^*)$ — the Hilbert spaces of holomorphic 1-forms on corresponding domains, are, respectively, naturally isomorphic to the Hilbert spaces $A_2^1(\mathbb{D})$ and $A_2^1(\mathbb{D}^*)$.

Consider generalized Faber polynomials of g and f defined, respectively, by [**Pom92, Teo03**]

$$\log \frac{g(z) - w}{bz} = -\sum_{n=1}^{\infty} \frac{P_n(w)}{n} z^{-n},$$

$$\log \frac{w - f(z)}{w} = \log \frac{f(z)}{z} - \sum_{n=1}^{\infty} \frac{Q_n(w)}{n} z^n.$$

Here $P_n(w)$ is a polynomial of degree n in w and $Q_n(w)$ is a polynomial of degree n in $1/w$. Specifically,

$$P_n(w) = (g^{-1}(w))_{\geq 0}^n,$$

the polynomial part of the n-th power of the inverse function g^{-1}, and

$$Q_n(w) = (f^{-1}(w))_{\leq 0}^{-n},$$

the principal part of the negative n-th power of the inverse function f^{-1}. Here for $S \subset \mathbb{Z}$ and a formal power series $A(w) = \sum_{n \in \mathbb{Z}} A_n w^n$ we denote $(A(w))_S = \sum_{n \in S} A_n w^n$.

Comparing the definition of Faber polynomials with the definition of Grunsky coefficients, we obtain the following relations (see, e.g. [**Pom92, Teo03**])

$$P_n(g(z)) = z^n + n\sum_{m=1}^{\infty} b_{nm} z^{-m}, \qquad P_n(f(z)) = nb_{n,0} + n\sum_{m=1}^{\infty} b_{n,-m} z^m,$$

$$Q_n(g(z)) = -nb_{-n,0} + n\sum_{m=1}^{\infty} b_{m,-n} z^{-m}, \qquad Q_n(f(z)) = z^{-n} + n\sum_{m=1}^{\infty} b_{-n,-m} z^m.$$

Now assume that the pair (f,g) is such that the corresponding set $F = \mathbb{C} \setminus \{f(\mathbb{D}) \cup g(\mathbb{D}^*)\}$ has Lebesgue measure zero. Then it follows from the above formulas and Remark 2.3 that the Hilbert spaces $A_2^1(\Omega)$ and $A_2^1(\Omega^*)$ have natural bases $\{\alpha_n\}_{n=1}^{\infty}$ and $\{\beta_n\}_{n=1}^{\infty}$, given respectively by the polynomials

$$\alpha_n(z) = \frac{P_n'(z)}{\sqrt{\pi n}}, \quad n \in \mathbb{N},$$

and by the Laurent polynomials

$$\beta_n(z) = \frac{Q_n'(z)}{\sqrt{\pi n}}, \quad n \in \mathbb{N}.$$

Indeed, we have

$$\alpha_n \circ f f' = \sum_{m=1}^{\infty} (B_3)_{nm} e_m \quad \text{and} \quad \beta_n \circ g\, g' = \sum_{m=1}^{\infty} (B_2)_{nm} f_m,$$

and the inner products are given by

$$\langle \alpha_n, \alpha_m \rangle = \iint_{\Omega} \alpha_n(z) \overline{\alpha_m(z)} d^2 z = \iint_{\mathbb{D}} \alpha_n(f(z)) f'(z) \overline{\alpha_m(f(z)) f'(z)} d^2 z$$

$$= \sum_{k=1}^{\infty} (B_3)_{nk} \overline{(B_3)}_{mk}.$$

Hence the period matrix of $A_2^1(\Omega)$ with respect to the basis $\{\alpha_n\}_{n=1}^{\infty}$ of holomorphic 1-forms on Ω (the Gram matrix of the basis) is given by

$$N_\Omega = \{\langle \alpha_n, \alpha_m \rangle\}_{m,n=1}^{\infty} = B_3 B_3^*.$$

Similarly, the period matrix of the basis $\{\beta_n\}_{n=1}^{\infty}$ of holomorphic 1-forms on Ω^* is given by

$$N_{\Omega^*} = \{\langle \beta_n, \beta_m \rangle\}_{m,n=1}^{\infty} = B_2 B_2^*.$$

We just proved the following result.

COROLLARY 2.16. *Let (f,g) be a normalized disjoint pair of univalent fuctions such that the set $F = \mathbb{C} \setminus \{f(\mathbb{D}) \cup g(\mathbb{D}^*)\}$ has Lebesgue measure zero and the corresponding Grunsky operators B_1 and B_4 are Hilbert-Schmidt. Then for $\mathcal{C} = f(S^1)$,*

$$\mathrm{Det}_F(\mathcal{C}) = \det N_\Omega = \det N_{\Omega^*}$$

3. Variations of the functions S_1 and S_2

Let ∂ and $\bar{\partial}$ be $(1,0)$ and $(0,1)$ components of de Rham differential d on the complex manifold $T_0(1)$. Here we compute the "first variations" of the functions S_1 and S_2 — the $(1,0)$-forms ∂S_1 and ∂S_2 on $T_0(1)$.

3.1. The first variation of S_2.

THEOREM 3.1. *The real-valued function $S_2 : T_0(1) \to \mathbb{R}$ is differentiable at every point $[\nu] \in T_0(1)$. In terms of the Bers coordinates ε_μ on the chart V_ν,*

$$\frac{\partial S_2}{\partial \varepsilon_\mu}([\nu]) = -\frac{1}{6\pi} \iint_{\mathbb{D}^*} \mathcal{S}(g_\nu)(z)\mu(z)d^2z.$$

Here $w_\nu = g_\nu^{-1} \circ f^\nu$ is the conformal welding corresponding to $[\nu] \in T_0(1)$.

PROOF. By definition of the Bers coordinates (see Section 2.3. in Chapter 1), for $\mu \in H^{-1,1}(\mathbb{D}^*)$

$$\frac{\partial S_2}{\partial \varepsilon_\mu}([\nu]) = \left.\frac{d}{d\varepsilon}\right|_{\varepsilon=0} S_2([\varepsilon\mu * \nu]).$$

Set $w_{\varepsilon\mu} \circ w_\nu = g_\varepsilon^{-1} \circ f^\varepsilon$, $f = f^0 = f^\nu$, $g = g_0 = g_\nu$ and $K_1(\varepsilon) = K_1(f^\varepsilon)$. Since $K_1(\varepsilon)$ is a holomorphic family, we have

$$(3.1) \qquad \frac{\partial S_2}{\partial \varepsilon_\mu}([\nu]) = \left.\frac{\partial}{\partial \varepsilon}\right|_{\varepsilon=0} \log \det(I - K_1(\varepsilon)) = -\operatorname{Tr}\left((I - K_1)^{-1}\frac{\partial K_1}{\partial \varepsilon}(0)\right)$$

(see, e.g., [**GK69**, Ch. IV.1, Property 9]). Now using Lemma 2.7, we have

$$\begin{aligned}\frac{\partial K_1}{\partial \varepsilon_\mu}([\nu])(z,w) &= \frac{1}{\pi^2} \iint_{\mathbb{D}} \iint_{\mathbb{D}^*} \frac{\mu(u)f'(z)g'(u)^2 f'(\zeta)}{(f(z)-g(u))^2(g(u)-f(\zeta))^2} K_1^*(\zeta,w) d^2u d^2\zeta \\ &= \iint_{\mathbb{D}} \iint_{\mathbb{D}^*} \mu(u) K_2(z,u) K_3(u,\zeta) K_1^*(\zeta,w) d^2u d^2\zeta \\ &= -\iint_{\mathbb{D}^*} \iint_{\mathbb{D}^*} \mu(u) K_2(z,u) K_4(u,\zeta) K_2^*(\zeta,w) d^2u d^2\zeta.\end{aligned}$$

Here in the last line, we have used the second relation in (2.2),

$$K_3 K_1^* = -K_4 K_2^*.$$

Let $R_2(z,w)$ be the kernel of the inverse operator K_2^{-1} — the anti-holomorphic function on $\mathbb{D}^* \times \mathbb{D}$ satisfying

$$\iint_{\mathbb{D}^*} K_2(z,\zeta)R_2(\zeta,w)d^2\zeta = I_1(z,w) = \frac{1}{\pi(1-z\bar{w})^2},$$

$$\iint_{\mathbb{D}} R_2(z,\zeta)K_2(\zeta,w)d^2\zeta = I_2(z,w) = \frac{1}{\pi(1-\bar{z}w)^2}.$$

Here $I_1(z,w)$ and $I_2(z,w)$ are the kernels of the identity operators on $A_2^1(\mathbb{D})$ and $\overline{A_2^1(\mathbb{D}^*)}$ respectively. Similarly, let $R_2^*(z,w)$ be the kernel of the inverse operator $(K_2^*)^{-1}$. We have

$$(I - K_1)^{-1} = K_2^{-1} = (K_2^*)^{-1} K_2^{-1},$$

so that

$$\frac{\partial S_2}{\partial \varepsilon_\mu}([\nu]) = \iint_{\mathbb{D}} \iint_{\mathbb{D}^*} \iint_{\mathbb{D}} \iint_{\mathbb{D}^*} \iint_{\mathbb{D}^*} \mu(u) R_2^*(w,\eta) R_2(\eta, z)$$
$$K_2(z,u) K_4(u,\zeta) K_2^*(\zeta, w) d^2 u d^2 \zeta d^2 z d^2 \eta d^2 w$$
$$= \iint_{\mathbb{D}^*} \iint_{\mathbb{D}^*} \iint_{\mathbb{D}^*} \mu(u) K_4(u,\zeta) I_2(\eta, u) I_2(\zeta, \eta) d^2 u d^2 \eta d^2 \zeta$$
$$= \iint_{\mathbb{D}^*} \iint_{\mathbb{D}^*} \mu(u) K_4(u,\zeta) I_2(\zeta, u) d^2 u d^2 \zeta$$
$$= \iint_{\mathbb{D}^*} \mu(u) K_4(u,u) d^2 u = -\frac{1}{6\pi} \iint_{\mathbb{D}^*} \mathcal{S}(g_\nu)(u) \mu(u) d^2 u.$$

Here in the last line we have used

$$K_4(u,u) = -\frac{1}{\pi} \lim_{\zeta \to u} \left(\frac{g'(u)g'(\zeta)}{(g(\zeta) - g(u))^2} - \frac{1}{(\zeta - u)^2} \right) = -\frac{1}{6\pi} \mathcal{S}(g)(u).$$

$\square$

Denote by $T^*_{[\mu]} T_0(1)$ the holomorphic cotangent space to $T_0(1)$ at a point $[\mu] \in T_0(1)$. The natural isomorphism $T_{[\mu]} T_0(1) \simeq H^{-1,1}(\mathbb{D}^*)$ induces the isomorphism $T^*_{[\mu]} T_0(1) \simeq A_2(\mathbb{D}^*)$. Define a holomorphic 1-form ϑ on $T_0(1)$ by

$$\vartheta_{[\mu]} = \mathcal{S}(g_\mu) \in A_2(\mathbb{D}^*),$$

where $w_\mu = g_\mu^{-1} \circ f^\mu \in T_0(1)$.

COROLLARY 3.2. *On* $T_0(1)$,

$$\partial \mathsf{S}_2 = -\frac{1}{6\pi} \vartheta.$$

REMARK 3.3. For C^3 curves the statement of Theorem 3.1 was obtained by Schiffer in [**Sch59**]. The derivation in [**Sch59**] uses the variational theory of Fredholm eigenvalues and the exterior variation of the domain. Our proof is different from Schiffer's: we use general formula (3.1) and the quasi-conformal variation.

3.2. The first variation of S_1. In addition to S_1, we introduce another function $\tilde{\mathsf{S}}_1 : T_0(1) \to \mathbb{R}$ defined by

$$\tilde{\mathsf{S}}_1([\mu]) = \mathsf{S}_1([\mu^{-1}]).$$

Using (1.3), we get

$$\tilde{\mathsf{S}}_1([\mu]) = \iint_{\mathbb{D}} \left| \mathcal{A}(f^\mu) - 2\frac{(f^\mu)'}{f^\mu} + \frac{2}{z} \right|^2 d^2 z + \iint_{\mathbb{D}^*} \left| \mathcal{A}(g_\mu) - 2\frac{g'_\mu}{g_\mu} + \frac{2}{z} \right|^2 d^2 z$$
$$- 4\pi \log |(g_\mu)'(\infty)|$$
$$= \iint_{\mathbb{D}} |\mathcal{A}(\tilde{g}_\mu)|^2 d^2 z + \iint_{\mathbb{D}^*} \left| \mathcal{A}(\tilde{f}^\mu) \right|^2 d^2 z + 4\pi \log |\tilde{g}'_\mu(0)|,$$

where $\tilde{f}^\mu = \imath \circ f^\mu \circ \imath$, $\tilde{g}_\mu = \imath \circ g_\mu \circ \imath$ and $\imath(z) = \frac{1}{z}$. The functions $\tilde{f}^\mu$ and $\tilde{g}_\mu$ are univalent, respectively, on the domains $\mathbb{D}^*$ and $\mathbb{D}$ and are normalized as $\tilde{f}^\mu(\infty) = \infty$, $(\tilde{f}^\mu)'(\infty) = 1$ and $\tilde{g}_\mu(0) = 0$. They satisfy the factorization

$$\tilde{w}_\mu = \tilde{g}_\mu^{-1} \circ \tilde{f}^\mu, \tag{3.2}$$

where $\tilde{w}_\mu = \imath \circ w_\mu \circ \imath$.

This description corresponds to the realization of $T(1)$ associated with the model $\mathbb{H}^2 \simeq \mathbb{D}$. Namely, due to the canonical isomorphism

$$\mu \in L^\infty(\mathbb{D}^*) \mapsto \tilde{\mu} = \imath^* \mu = \mu\left(\frac{1}{z}\right) \frac{z^2}{\bar{z}^2} \in L^\infty(\mathbb{D}),$$

we have $T(1) \simeq L^\infty(\mathbb{D})_1/\sim$. If w_μ is a q.c. mapping associated with $\mu \in L^\infty(\mathbb{D}^*)_1$, then $\tilde{w}_\mu$ is the q.c. mapping associated with $\tilde{\mu} \in L^\infty(\mathbb{D})_1$, and corresponding conformal wielding is given by (3.2).

In this section, we will also use the model $T(1) \simeq L^\infty(\mathbb{D})_1$. To simplify the notations, for $\mu \in L^\infty(\mathbb{D})_1$ we will denote corresponding q.c. mapping by $w_\mu = g_\mu^{-1} \circ f^\mu$, where f^μ and g_μ are univalent on the domains $\mathbb{D}^*$ and $\mathbb{D}$ and are normalized as $f^\mu(\infty) = \infty$, $(f^\mu)'(\infty) = 1$ and $g_\mu(0) = 0$. Correspondingly, for $\gamma = g^{-1} \circ f \in T(1)$ we would have the normalization $f(\infty) = \infty$, $f'(\infty) = 1$ and $g(0) = 0$. To avoid confusion with the notations for our primary model $T(1) = L^\infty(\mathbb{D}^*)_1/\sim$, we will always specify explicitly in the main text when we are using the model $T(1) \simeq L^\infty(\mathbb{D})_1/\sim$.

The function S_1 on $\mathcal{T}_0(1)$ naturally extends to a function $\hat{\mathsf{S}}$ on $\mathcal{T}_0(1)$, defined by

$$\hat{\mathsf{S}}(\gamma) = \iint_{\mathbb{D}} |\mathcal{A}(f)|^2 \, d^2z + \iint_{\mathbb{D}^*} |\mathcal{A}(g)|^2 \, d^2z - 4\pi \log|g'(\infty)|,$$

where $\gamma = g^{-1} \circ f \in \mathcal{T}_0(1)$. For $\tilde{\mathsf{S}}(\gamma) = \hat{\mathsf{S}}(\gamma^{-1})$ we have

$$\tilde{\mathsf{S}}(\gamma) = \iint_{\mathbb{D}} |\mathcal{A}(\tilde{g})|^2 \, d^2z + \iint_{\mathbb{D}^*} \left|\mathcal{A}(\tilde{f})\right|^2 d^2z + 4\pi \log|\tilde{g}'(0)|,$$

where $\tilde{f} = \imath \circ f \circ \imath$ and $\tilde{g} = \imath \circ g \circ \imath$.

LEMMA 3.4. *The function $\tilde{\mathsf{S}}$ is constant along the fibers of the canonical projection $\pi : \mathcal{T}_0(1) \to T_0(1)$, $\tilde{\mathsf{S}} = \tilde{\mathsf{S}}_1 \circ \pi$.*

PROOF. We are using the model $T(1) \simeq L^\infty(\mathbb{D})_1/\sim$. For $\mu \in L^\infty(\mathbb{D})_1$ let $\gamma = g^{-1} \circ f$, $\gamma_\mu = g_\mu^{-1} \circ f^\mu \in \mathcal{T}_0(1)$ be such that $\pi(\gamma) = \pi(\gamma_\mu) = [\mu]$. Comparing the normalization for f and f^μ at ∞, we get

$$f = \sigma \circ f^\mu \quad \text{and} \quad g = \sigma \circ g_\mu \circ \alpha^{-1},$$

for some $\alpha \in \mathrm{PSU}(1,1)$ and $\sigma(z) = z + b_0$. Since $f \mapsto \mathcal{A}(f)$ is invariant if f is post-composed with a translation[1], to prove that $\tilde{\mathsf{S}}(\gamma) = \tilde{\mathsf{S}}(\gamma_\mu)$ we need only to check that for $\alpha \in \mathrm{PSU}(1,1)$,

$$\iint_{\mathbb{D}} |\mathcal{A}(g \circ \alpha^{-1})|^2 d^2z + 4\pi \log|(g \circ \alpha^{-1})'(0)| = \iint_{\mathbb{D}} |\mathcal{A}(g)|^2 d^2z + 4\pi \log|g'(0)|.$$

---

[1]This is why it is more convenient to use the model $T(1) \simeq L^\infty(\mathbb{D})_1/\sim$.

Let
$$\alpha(z) = e^{i\theta}\frac{z-w}{1-z\bar{w}}$$
and set $\log g'(z) = \sum_{n=0}^{\infty} a_n z^n$. Then $\mathcal{A}(g) = \sum_{n=1}^{\infty} n a_n z^{n-1}$ and
$$\iint_{\mathbb{D}} |\mathcal{A}(g\circ\alpha^{-1})|^2 d^2z = \iint_{\mathbb{D}} |\mathcal{A}(g)\circ\alpha^{-1}(\alpha^{-1})' + \mathcal{A}(\alpha^{-1})|^2 d^2z$$
$$= \iint_{\mathbb{D}} |\mathcal{A}(g) - \mathcal{A}(\alpha)|^2 d^2z = \iint_{\mathbb{D}} \left|\mathcal{A}(g) - \frac{2\bar{w}}{1-z\bar{w}}\right|^2 d^2z$$
$$= \iint_{\mathbb{D}} |\mathcal{A}(g)|^2 d^2z - 4\operatorname{Re}\left(w\iint_{\mathbb{D}} \mathcal{A}(g)(z)\sum_{n=1}^{\infty}(w\bar{z})^{n-1} d^2z\right)$$
$$+ 4|w|^2 \iint_{\mathbb{D}} \left|\sum_{n=1}^{\infty}(w\bar{z})^{n-1}\right|^2 d^2z.$$

The last two terms give
$$-4\pi\operatorname{Re}\left(\sum_{n=1}^{\infty} a_n w^n\right) + 4\pi\sum_{n=1}^{\infty}\frac{|w|^{2n}}{n}$$
$$= -4\pi\log|g'(w)| + 4\pi\log|g'(0)| - 4\pi\log(1-|w|^2).$$

On the other hand, we have
$$(g\circ\alpha^{-1})'(0) = g'(\alpha^{-1}(0))(\alpha^{-1})'(0) = (1-|w|^2)g'(w).$$

This concludes the proof. □

THEOREM 3.5. *The real-valued function* $\tilde{S}_1 : T_0(1) \to \mathbb{R}$ *is differentiable at every point* $[\nu] \in T_0(1)$. *In terms of the Bers coordinates* ε_μ *on the chart* V_ν,
$$\frac{\partial \tilde{S}_1}{\partial \varepsilon_\mu}([\nu]) = 2\iint_{\mathbb{D}^*} \mathcal{S}(g_\nu)(z)\mu(z) d^2z.$$

PROOF. We are using the model $T(1) \simeq L^\infty(\mathbb{D})_1$. For $[\nu] \in T_0(1)$ choose a representative $\nu \in L^\infty(\mathbb{D})_1$ which is a product of elements in $H^{-1,1}(\mathbb{D})_1$, and let $w_\varepsilon = w_{\varepsilon\mu}\circ w_\nu = g_\varepsilon^{-1}\circ f^\varepsilon$. It follows from Lemma 1.5 in Chapter 1 that corresponding $\gamma_\varepsilon = g_\varepsilon^{-1}\circ f^\varepsilon$ fixes $0, 1, \infty$. By the above lemma,
$$\tilde{S}_1([\varepsilon\mu * \nu]) = \tilde{S}(\gamma_\varepsilon).$$
We have $\gamma_\varepsilon \circ \gamma_\nu^{-1} = \gamma_{\varepsilon\kappa}$, where $\kappa = (\alpha^{-1})^*(\mu)$ and $\alpha = \gamma_\nu \circ w_\nu^{-1} \in \mathrm{PSU}(1,1)$. Set $f = f^0$, $g = g^0$, so that $g = \sigma \circ g_\nu \circ \alpha^{-1}$ for some $\sigma \in \mathrm{PSL}(2,\mathbb{C})$, and define $v_\varepsilon = f^\varepsilon \circ f^{-1}$. Since f^ε is normalized, it's Laurent expansion at ∞ has the form
$$f^\varepsilon(z) = z + \frac{b_1}{z} + \frac{b_2}{z^2} + \ldots.$$

Hence
$$\left.\frac{\partial}{\partial\varepsilon}\right|_{\varepsilon=0} v_\varepsilon(z) = O(z^{-1}) \quad \text{as} \quad z \to \infty,$$

and the first variations of v_ε have the form

$$\frac{\partial}{\partial \varepsilon}\bigg|_{\varepsilon=0} v_\varepsilon(z) = -\frac{1}{\pi} \iint_\Omega \frac{((g^{-1})^*\kappa)(w)}{w-z} d^2w, \quad \frac{\partial}{\partial \bar\varepsilon}\bigg|_{\varepsilon=0} v_\varepsilon(z) = 0,$$

where $\Omega = g(\mathbb{D})$. Since $\gamma_{\varepsilon\kappa}$ fixes $0, 1, \infty$, we also have

$$\frac{\partial}{\partial \varepsilon}\bigg|_{\varepsilon=0} \gamma_{\varepsilon\kappa}(z) = -\frac{1}{\pi} \iint_\mathbb{D} \frac{z(z-1)\kappa(w)}{(w-z)w(w-1)} d^2w,$$

$$\frac{\partial}{\partial \bar\varepsilon}\bigg|_{\varepsilon=0} \gamma_{\varepsilon\kappa}(z) = -\frac{1}{\pi} \iint_\mathbb{D} \frac{z(z-1)\overline{\kappa(w)}}{(1-\bar w z)\bar w(1-\bar w)} d^2w.$$

Using $f^\varepsilon = v_\varepsilon \circ f$, we obtain

$$\mathcal{A}(f^\varepsilon) = \mathcal{A}(v_\varepsilon) \circ f f' + \mathcal{A}(f).$$

Applying the variational formulas for v_ε, we have

$$\frac{\partial}{\partial \varepsilon}\bigg|_{\varepsilon=0} \mathcal{A}(v_\varepsilon)(z) = \frac{\partial^2}{\partial z^2} \frac{\partial}{\partial \varepsilon}\bigg|_{\varepsilon=0} v_\varepsilon(z) = -\frac{2}{\pi} \iint_\Omega \frac{((g^{-1})^*\kappa)(w)}{(w-z)^3} d^2w,$$

and hence

$$\frac{\partial}{\partial \varepsilon}\bigg|_{\varepsilon=0} \iint_{\mathbb{D}^*} |\mathcal{A}(f^\varepsilon)|^2 d^2z = -\frac{2}{\pi} \iint_{\mathbb{D}^*} \iint_\mathbb{D} \frac{\kappa(w)g'(w)^2 f'(z)}{(g(w)-f(z))^3} \overline{\mathcal{A}(f)(z)} d^2w d^2z = I_1.$$

Similarly, using

$$g_\varepsilon \circ \gamma_{\varepsilon\kappa} = v_\varepsilon \circ g,$$

we have

$$g'_\varepsilon \circ \gamma_{\varepsilon\kappa} (\gamma_{\varepsilon\kappa})_z = v'_\varepsilon \circ g \, g',$$

and

$$\mathcal{A}(g_\varepsilon) \circ \gamma_{\varepsilon\kappa}(\gamma_{\varepsilon\kappa})_z + \mathcal{A}(\gamma_{\varepsilon\kappa}) = \mathcal{A}(v_\varepsilon) \circ g \, g' + \mathcal{A}(g),$$

where

$$\mathcal{A}(\gamma_{\varepsilon\kappa}) = \frac{(\gamma_{\varepsilon\kappa})_{zz}}{(\gamma_{\varepsilon\kappa})_z}.$$

Hence we have

$$\frac{\partial}{\partial \varepsilon}\bigg|_{\varepsilon=0} g'_\varepsilon(0) = -g''(0)\left(\frac{\partial}{\partial \varepsilon}\bigg|_{\varepsilon=0} \gamma_{\varepsilon\kappa}\right)(0) - g'(0)\frac{\partial}{\partial z}\left(\frac{\partial}{\partial \varepsilon}\bigg|_{\varepsilon=0} \gamma_{\varepsilon\kappa}\right)(0)$$

$$+ g'(0)\frac{\partial}{\partial z}\left(\frac{\partial}{\partial \varepsilon}\bigg|_{\varepsilon=0} v_\varepsilon\right)(0)$$

$$= \frac{g'(0)}{\pi} \iint_\mathbb{D} \kappa(w)\left(\frac{1}{w^2} - \frac{1}{w(w-1)} - \frac{g'(w)^2}{g(w)^2}\right) d^2w,$$

and
$$\left.\frac{\partial}{\partial\varepsilon}\right|_{\varepsilon=0}\overline{g'_\varepsilon(0)} = -\overline{g''(0)}\overline{\left(\left.\frac{\partial}{\partial\bar\varepsilon}\right|_{\varepsilon=0}\gamma_{\varepsilon\kappa}\right)(0)} - \overline{g'(0)}\overline{\frac{\partial}{\partial z}\left(\left.\frac{\partial}{\partial\bar\varepsilon}\right|_{\varepsilon=0}\gamma_{\varepsilon\kappa}\right)(0)}$$
$$= \frac{\overline{g'(0)}}{\pi}\iint_{\mathbb{D}}\frac{\kappa(w)}{w(w-1)}d^2w,$$

as well as
$$\left.\frac{\partial}{\partial\varepsilon}\right|_{\varepsilon=0}\mathcal{A}(g_\varepsilon)\circ\gamma_{\varepsilon\kappa}(\gamma_{\varepsilon\kappa})_z = \left(\left.\frac{\partial}{\partial\varepsilon}\right|_{\varepsilon=0}\mathcal{A}(v_\varepsilon)\right)\circ gg' - \left.\frac{\partial}{\partial\varepsilon}\right|_{\varepsilon=0}\mathcal{A}(\gamma_{\varepsilon\kappa})$$
$$= -\frac{2}{\pi}\iint_{\mathbb{D}}\kappa(w)\left(\frac{g'(w)^2 g'(z)}{(g(w)-g(z))^3} - \frac{1}{(w-z)^3}\right)d^2w,$$

and
$$\left.\frac{\partial}{\partial\varepsilon}\right|_{\varepsilon=0}\overline{\mathcal{A}(g_\varepsilon)\circ\gamma_{\varepsilon\kappa}(\gamma_{\varepsilon\kappa})_z} = -\left.\frac{\partial}{\partial\bar\varepsilon}\right|_{\varepsilon=0}\overline{\mathcal{A}(\gamma_{\varepsilon\kappa})} = \frac{2}{\pi}\iint_{\mathbb{D}}\frac{\kappa(w)}{(1-w\bar z)^3 w}d^2w.$$

From here we get
$$2\pi\left.\frac{\partial}{\partial\varepsilon}\right|_{\varepsilon=0}\log|g'_\varepsilon(0)|^2 = -2\iint_{\mathbb{D}}\kappa(w)\left(\frac{g'(w)^2}{g(w)^2} - \frac{1}{w^2}\right)d^2w = I_2,$$

and
$$\left.\frac{\partial}{\partial\varepsilon}\right|_{\varepsilon=0}\iint_{\mathbb{D}}|\mathcal{A}(g_\varepsilon)|^2 d^2z = \left.\frac{\partial}{\partial\varepsilon}\right|_{\varepsilon=0}\iint_{\mathbb{D}}|\mathcal{A}(g_\varepsilon)\circ\gamma_{\varepsilon\kappa}(\gamma_{\varepsilon\kappa})_z|^2(1-|\varepsilon\kappa|^2)d^2z$$
$$= -\frac{2}{\pi}\iint_{\mathbb{D}}\iint_{\mathbb{D}}\kappa(w)\left(\frac{g'(w)^2 g'(z)}{(g(w)-g(z))^3} - \frac{1}{(w-z)^3}\right)\overline{\mathcal{A}(g)(z)}d^2w d^2z$$
$$+ \frac{2}{\pi}\iint_{\mathbb{D}}\iint_{\mathbb{D}}\frac{\kappa(w)\overline{\mathcal{A}(g)(z)}}{w(1-w\bar z)^3}d^2w d^2z = I_3 + I_4.$$

Let $\log g'(z) = \sum_{n=0}^\infty a_n z^n$ be the power series expansion of $\log g'(z)$. Then $\mathcal{A}(g) = \sum_{n=1}^\infty n a_n z^{n-1}$. Explicit computation gives
$$\frac{2}{\pi}\iint_{\mathbb{D}}\frac{\overline{\mathcal{A}(g)(z)}}{w(1-w\bar z)^3}d^2z = \sum_{n=1}^\infty n(n+1)\overline{a_n}w^{n-2} = \overline{\mathcal{A}(g)}'(w) + \frac{2}{w}\overline{\mathcal{A}(g)}(w).$$

Hence
$$I_4 = \iint_{\mathbb{D}}\kappa(w)\left(\overline{\mathcal{A}(g)}'(w) + \frac{2}{w}\overline{\mathcal{A}(g)}(w)\right)d^2w.$$

To compute the other terms, we define the following holomorphic function on $\mathbb{D}$,
$$h(w) = \frac{1}{\pi}\iint_{\mathbb{D}^*}\frac{g'(w)f'(z)}{(g(w)-f(z))^2}\overline{\mathcal{A}(f)(z)}d^2z$$
$$+ \frac{1}{\pi}\iint_{\mathbb{D}}\left(\frac{g'(w)g'(z)}{(g(w)-g(z))^2} - \frac{1}{(w-z)^2}\right)\overline{\mathcal{A}(g)(z)}d^2z.$$

Then it is easy to check that
$$\left.\frac{\partial}{\partial \varepsilon}\right|_{\varepsilon=0} \tilde{\mathsf{S}}(\gamma_\varepsilon) = I_1 + I_2 + I_3 + I_4$$
$$= \iint_{\mathbb{D}} \kappa(w)\left(h'(w) - \mathcal{A}(g)(w)h(w) - 2\frac{g'(w)^2}{g(w)^2} + \frac{2}{w^2} + \mathcal{A}(g)'(w) + \frac{2}{w}\mathcal{A}(g)(w)\right) d^2w.$$

To finish the proof, we claim that
$$h(w) = \mathcal{A}(g)(w) - 2\frac{g'(w)}{g(w)} + \frac{2}{w},$$
which is going to be proved in the next lemma. With this equation for h, it is straightforward to compute that
$$\left.\frac{\partial}{\partial \varepsilon}\right|_{\varepsilon=0} \tilde{\mathsf{S}}(\gamma_\varepsilon) = \iint_{\mathbb{D}} \left(2\mathcal{A}(g)'(w) - \mathcal{A}(g)(w)^2\right) \kappa(w) d^2w$$
$$= 2\iint_{\mathbb{D}} \mathcal{S}(g)(w)\kappa(w) d^2w = 2\iint_{\mathbb{D}} \mathcal{S}(g_\nu)(w)\mu(w) d^2w.$$

Returning back to the model $T(1) = L^\infty(\mathbb{D}^*)_1/\sim$, we get the statement of the theorem. $\square$

LEMMA 3.6. *In the model* $T(1) \simeq L^\infty(\mathbb{D})_1/\sim$, *let* $\gamma = g^{-1} \circ f$ *be the conformal welding corresponding to* $\gamma \in T_0(1)$. *Then for* $z \in \mathbb{D}$ *the following identity holds*
$$\mathcal{A}(g)(z) - 2\frac{g'(z)}{g(z)} + \frac{2}{z} = \frac{1}{\pi} \iint_{\mathbb{D}^*} \frac{g'(z)f'(w)}{(g(z)-f(w))^2}\overline{\mathcal{A}(f)(w)} d^2w$$
$$+ \frac{1}{\pi} \iint_{\mathbb{D}} \left(\frac{g'(z)g'(w)}{(g(w)-g(w))^2} - \frac{1}{(z-w)^2}\right) \overline{\mathcal{A}(g)(w)} d^2w.$$

PROOF. First we consider the case when $\mathcal{A}(g)$ and $\mathcal{A}(f)$ are smooth functions on S^1. Specifically, we assume that the Beltrami differential μ corresponding to $\pi(\gamma) \in T_0(1)$, is smooth on $\mathbb{C}$ and $\mu|_{S^1} = \mu_{\bar{z}}|_{S^1} = 0$. Denote by $h(z)$ the right-hand side of the identity of the lemma. Changing the variables of integration and using Stokes' theorem, we obtain
$$h \circ g^{-1}(g^{-1})'(z) = -\frac{1}{\pi} \iint_{\Omega^*} \frac{\overline{\mathcal{A}(f^{-1})(w)}}{(z-w)^2} d^2w - \frac{1}{\pi} \iint_{\Omega} \frac{\overline{\mathcal{A}(g^{-1})(w)}}{(z-w)^2} d^2w$$
$$= \frac{1}{2\pi i} \oint_\mathcal{C} \frac{1}{(z-w)} \left(\overline{\mathcal{A}(g^{-1})(w)} - \overline{\mathcal{A}(f^{-1})(w)}\right) d\bar{w},$$
where $\Omega = g(\mathbb{D}), \Omega^* = f(\mathbb{D}^*)$ and $\mathcal{C} = g(S^1)$. Next, consider the relation $\tilde{\gamma} \circ g^{-1} = f^{-1}$, where $\tilde{\gamma} = \gamma^{-1}$, and differentiate it twice with respect to z. Since $\tilde{\gamma}_{\bar{z}}$ vanishes on S^1, we get the following relations on $\mathcal{C}$,
$$\frac{\tilde{\gamma}_z}{\tilde{\gamma}} \circ g^{-1}(g^{-1})_z = \frac{(f^{-1})_z}{f^{-1}},$$
$$\mathcal{A}(\tilde{\gamma}) \circ g^{-1}(g^{-1})_z = \mathcal{A}(f^{-1}) - \mathcal{A}(g^{-1}).$$

Hence
$$h \circ g^{-1}(g^{-1})'(z) = -\frac{1}{2\pi i} \oint_C \frac{1}{(z-w)} \overline{(\mathcal{A}(\tilde\gamma) \circ g^{-1})(w)(g^{-1})_w(w)} d\bar w$$
$$= -\frac{1}{2\pi i} \oint_{S^1} \frac{1}{z-g(w)} \overline{\mathcal{A}(\tilde\gamma)(w)} d\bar w.$$

On the other hand, since $j \circ \tilde\gamma = \tilde\gamma \circ j$, where j is the inversion $z \mapsto \frac{1}{z}$, we have
$$\overline{\mathcal{A}(\tilde\gamma)} = \mathcal{A}(\tilde\gamma) \circ j \, j_{\bar z} - 2\frac{\tilde\gamma_z}{\tilde\gamma} \circ j \, j_{\bar z} + \overline{\mathcal{A}(\tilde j)}.$$

Hence
$$h \circ g^{-1}(g^{-1})'(z) = \frac{1}{2\pi i} \oint_{S^1} \frac{1}{z-g(w)} \left(\mathcal{A}(\tilde\gamma)\left(\frac{1}{\bar w}\right) \frac{1}{\bar w^2} - 2\frac{\tilde\gamma_w}{\tilde\gamma}\left(\frac{1}{\bar w}\right) \frac{1}{\bar w^2} + \frac{2}{\bar w} \right) d\bar w$$
$$= -\frac{1}{2\pi i} \oint_{S^1} \frac{1}{z-g(w)} \left(\mathcal{A}(\tilde\gamma)(w) - 2\frac{\tilde\gamma_w(w)}{\tilde\gamma(w)} + \frac{2}{w} \right) dw$$
$$= \frac{1}{2\pi i} \oint_C \frac{1}{(z-w)} \left(\mathcal{A}(g^{-1})(w) - \mathcal{A}(f^{-1})(w) + 2\frac{(f^{-1})_w(w)}{f^{-1}(w)} - 2\frac{(g^{-1})_w(w)}{g^{-1}(w)} \right) dw.$$

The functions
$$\mathcal{A}(f^{-1})(z) - 2\frac{(f^{-1})_z(z)}{f^{-1}(z)} + \frac{2}{z} \quad \text{and} \quad \mathcal{A}(g^{-1})(z) - 2\frac{(g^{-1})_z(z)}{g^{-1}(z)} + \frac{2}{z}$$
are holomorphic on $\Omega^* = f(\mathbb{D}^*)$ and $\Omega = g(\mathbb{D})$ respectively and due to the normalization of f,
$$\mathcal{A}(f^{-1})(z) - 2\frac{(f^{-1})_z(z)}{f^{-1}(z)} + \frac{2}{z} = O\left(\frac{1}{z^2}\right) \quad \text{as} \quad z \to \infty.$$

Thus we have by Cauchy formula
$$h \circ g^{-1}(g^{-1})'(z) = -\left(\mathcal{A}(g^{-1})(z) - 2\frac{(g^{-1})'(z)}{g^{-1}(z)} + \frac{2}{z} \right)$$
or equivalently,
$$h(z) = \mathcal{A}(g)(z) - 2\frac{g'(z)}{g(z)} + \frac{2}{z}.$$

For a general point $\gamma = g^{-1} \circ f$ in $\mathcal{T}_0(1)$, we let $f_n = r_n^{-1} \circ f \circ r_n$, where r_n is the dilation $z \mapsto \frac{n+1}{n} z$. Since f_n is a normalized univalent function on $|z| > \frac{n}{n+1}$, corresponding $\gamma_n^{-1} = g_n^{-1} \circ f_n \in \mathcal{T}_0(1)$ satisfies the assumptions made in the beginning of the proof. Since $\mathcal{A}(f) \in A_2^1(\mathbb{D}^*)$, we see that
$$\|\mathcal{A}(\iota \circ f_n \circ \iota) - \mathcal{A}(\iota \circ f \circ \iota)\|_{A_2^1(\mathbb{D})}$$
$$= \left\| \left(\mathcal{A}(f_n) - 2\frac{f_n'}{f_n} + \frac{2}{z} \right) - \left(\mathcal{A}(f) - 2\frac{f'}{f} + \frac{2}{z} \right) \right\|_{A_2^1(\mathbb{D}^*)} \to 0 \quad \text{as} \quad n \to \infty.$$

By Corollary A.4 and Corollary A.6 in Appendix A we also have
$$\lim_{n\to\infty} \|\mathcal{A}(g_n) - \mathcal{A}(g)\|_{A_2^1(\mathbb{D})} = 0$$
and
$$\lim_{n\to\infty} \left\| \left(\mathcal{A}(g_n) - 2\frac{g_n'}{g_n} + \frac{2}{z} \right) - \left(\mathcal{A}(g) - 2\frac{g'}{g} + \frac{2}{z} \right) \right\|_{A_2^1(\mathbb{D})} = 0.$$

In particular, since convergence in $A_2^1(\mathbb{D})$ implies convergence in $A_\infty^1(\mathbb{D})$, we get
$$\lim_{n\to\infty}\left(\mathcal{A}(g_n)(z) - 2\frac{g_n'}{g_n}(z) + \frac{2}{z}\right) = \mathcal{A}(g)(z) - 2\frac{g'}{g}(z) + \frac{2}{z},$$
uniformly on compact subsets of $\mathbb{D}$. Since we have already shown that
$$h_n(z) = \mathcal{A}(g_n)(z) - 2\frac{g_n'}{g_n}(z) + \frac{2}{z},$$
to finish the proof of the lemma we need to verify that $\lim_{n\to\infty} h_n(z) = h(z)$ uniformly on compact subsets of $\mathbb{D}$.

We denote by $K_1[n]$ and $K_2[n]$ the operators associated with the disjoint pair of univalent functions (g_n, f_n), and by K_1 and K_2 — the operators associated with the pair (g, f). Then
$$h_n(z) - h(z) = -\left(K_1[n]\overline{\mathcal{A}(g_n)}\right)(z) + \left(K_1\overline{\mathcal{A}(g)}\right)(z)$$
$$+ \left(K_2[n]\overline{\mathcal{A}(f_n)}\right)(z) - \left(K_2\overline{\mathcal{A}(f)}\right)(z).$$

Now using Theorem B.1 from Appendix B, and the fact that the inverse map is continuous on $T_0(1)$, we get that
$$\lim_{n\to\infty}\|K_1[n] - K_1\| = 0,$$
where $\|\ \|$ stands for the norm of the Banach space $\mathscr{B}(\overline{A_2^1(\mathbb{D})}, A_2^1(\mathbb{D}))$. Since $\|K_1[n]\| \leq 1$, we have
$$\left\|K_1[n]\overline{\mathcal{A}(g_n)} - K_1\overline{\mathcal{A}(g)}\right\|_{A_2^1(\mathbb{D})}$$
$$\leq \left\|K_1[n]\overline{(\mathcal{A}(g_n) - \mathcal{A}(g))}\right\|_{A_2^1(\mathbb{D})} + \left\|(K_1[n] - K_1)\overline{\mathcal{A}(g)}\right\|_{A_2^1(\mathbb{D})}$$
$$\leq \|\mathcal{A}(g_n) - \mathcal{A}(g)\|_{A_2^1(\mathbb{D})} + \|K_1[n] - K_1\|\|\mathcal{A}(g)\|_{A_2^1(\mathbb{D})},$$
which tends to 0 as $n \to \infty$. Consequently,
$$\lim_{n\to\infty}(K_1[n]\overline{\mathcal{A}(g_n)})(z) = (K_1\overline{\mathcal{A}(g)})(z),$$
uniformly on compact subsets of $\mathbb{D}$.

To prove the convergence of the other term in $h_n(z) - h(z)$, we let $\mathfrak{g}_n = r_n \circ g_n$ and $\mathfrak{f}_n = r_n \circ f_n = f \circ r_n$. Let $\Omega_n^* = \mathfrak{f}_n(\mathbb{D}^*) = f(\{|z| > \frac{n+1}{n}\})$. Since $\Omega_n^* \subseteq \Omega_{n+1}^*$, the sequence of domains $\Omega_n = \mathfrak{g}_n(\mathbb{D})$ is a decreasing sequence that contains 0 and $\bigcap \mathfrak{g}_n(\mathbb{D}) = g(\mathbb{D}) = \Omega$. By Caratheodory kernel theorem (see, e.g., [**Pom92**]), the sequence of univalent functions $\mathfrak{g}_n : \mathbb{D} \to \mathbb{C}$ converges uniformly on compact sets to the univalent function $g : \mathbb{D} \to \mathbb{C}$. By Weierstrass theorem, $\lim_{n\to\infty} \mathfrak{g}_n'(z) = g'(z)$, uniformly on compact subsets of $\mathbb{D}$. Using that the operator K_2 is unaffected by a simultaneous post-composition of f and g with $\alpha \in \mathrm{PSL}(2,\mathbb{C})$ and that $\mathcal{A}(\mathfrak{f}_n) = \mathcal{A}(r_n \circ f_n) = \mathcal{A}(f_n)$, we have
$$\left(K_2[n]\overline{\mathcal{A}(f_n)}\right)(z) - \left(K_2\overline{\mathcal{A}(f)}\right)(z)$$
$$= \frac{1}{\pi}\iint_{\mathbb{D}^*} \frac{\mathfrak{g}_n'(z)\mathfrak{f}_n'(w)}{(\mathfrak{g}_n(z) - \mathfrak{f}_n(w))^2}\overline{\mathcal{A}(\mathfrak{f}_n)(w)}d^2w - \iint_{\mathbb{D}^*} \frac{g'(z)f'(w)}{(g(z) - f(w))^2}\overline{\mathcal{A}(f)(w)}d^2w$$
$$= u_n(\mathfrak{g}_n(z))\mathfrak{g}_n'(z) - u(g(z))g'(z).$$

3. VARIATIONS OF THE FUNCTIONS S_1 AND S_2

Here for $z \in \Omega_n = \mathfrak{g}_n(\mathbb{D})$ we set

$$u_n(z) = \frac{1}{\pi} \iint_{\mathbb{D}^*} \frac{\mathfrak{f}'_n(w)}{(z - \mathfrak{f}_n(w))^2} \overline{\mathcal{A}(\mathfrak{f}_n)(w)} d^2w = -\frac{1}{\pi} \iint_{\Omega_n^*} \frac{\overline{\mathcal{A}(\mathfrak{f}_n^{-1})(w)}}{(z - w)^2} d^2w,$$

and for $z \in \Omega = g(\mathbb{D})$,

$$u(z) = \frac{1}{\pi} \iint_{\mathbb{D}} \frac{f'(w)}{(z - f(w))^2} \overline{\mathcal{A}(f)(w)} d^2w = -\frac{1}{\pi} \iint_{\Omega^*} \frac{\overline{\mathcal{A}(f^{-1})(w)}}{(z - w)^2} d^2w.$$

Let $\tilde{u}_n = u_n|_\Omega$. Using $\mathcal{A}(\mathfrak{f}_n^{-1}) = \mathcal{A}(r_n^{-1} \circ f^{-1}) = \mathcal{A}(f^{-1})$ and $\Omega_n^* \subset \Omega^*$, we get

$$\tilde{u}_n(z) - u(z) = \frac{1}{\pi} \iint_{\Omega^* \setminus \Omega_n^*} \frac{\overline{\mathcal{A}(f^{-1})(w)}}{(z - w)^2} d^2w.$$

Since Hilbert transform is an isometry, we obtain

$$\|\tilde{u}_n \circ g \, g' - u \circ g \, g'\|^2_{A_2^1(\mathbb{D})} = \iint_{\mathbb{D}} |\tilde{u}_n(g(z))g'(z) - u(g(z))g'(z)|^2 \, d^2z$$

$$= \iint_{\Omega} |\tilde{u}_n(z) - u(z)|^2 d^2z \le \iint_{\Omega^* \setminus \Omega_n^*} |\mathcal{A}(f^{-1})(z)|^2 d^2z = \iint_{1 < |z| < \frac{n+1}{n}} |\mathcal{A}(f)(z)|^2 d^2z.$$

Since $\mathcal{A}(f) \in A_2^1(\mathbb{D}^*)$, we get

$$\lim_{n \to \infty} \|\tilde{u}_n \circ g \, g' - u \circ g \, g'\|_{A_2^1(\mathbb{D})} = 0,$$

and, consequently, $\lim_{n \to \infty} \tilde{u}_n(z) = u(z)$, uniformly on compact subsets of Ω. For every compact subset $E \subset \mathbb{D}$, $\mathfrak{g}_n(E) \subset \Omega$ for n sufficiently large, and it follows that

$$\lim_{n \to \infty} u_n(\mathfrak{g}_n(z))\mathfrak{g}'_n(z) = u(g(z))g'(z),$$

uniformly on E. $\square$

COROLLARY 3.7. *On $T_0(1)$,*

$$\partial \tilde{S}_1 = 2\vartheta.$$

THEOREM 3.8. *The functions $S_1, \tilde{S}_1, S_2$ on $T_0(1)$ satisfy the following relations,*

$$S_2 = -\frac{1}{12\pi} S_1 = -\frac{1}{12\pi} \tilde{S}_1.$$

In particular, in Bers coordinates ε_μ on the chart V_ν at $[\nu] \in T_0(1)$,

$$\frac{\partial S_1}{\partial \varepsilon_\mu}([\nu]) = 2 \iint_{\mathbb{D}^*} \mathcal{S}(g_\nu)(z) \mu(z) d^2z,$$

where $w_\nu = g_\nu^{-1} \circ f^\nu$ is the conformal welding corresponding to $[\nu] \in T_0(1)$.

PROOF. Since $S_2(0) = \tilde{S}_1(0) = 0$, Theorems 3.1 and 3.5 immediately give

$$S_2 = -\frac{1}{12\pi} \tilde{S}_1.$$

Since the function S_2 is symmetric,

$$S_2([\mu]) = S_2([\mu]^{-1}),$$

the function $\tilde{\mathsf{S}}_1$ is also symmetric, so that $\tilde{\mathsf{S}}_1 = \mathsf{S}_1$. $\square$

COROLLARY 3.9. *On $T_0(1)$,*
$$\partial \mathsf{S}_1 = 2\vartheta.$$

REMARK 3.10. Returning to the model $T(1) = L^\infty(\mathbb{D}^*)/\sim$, let $\gamma = g^{-1} \circ f \in T_0(1)$. Introducing the operator $\mathbf{K} : \overline{A_2^1(\mathbb{D})} \oplus \overline{A_2^1(\mathbb{D}^*)} \to A_2^1(\mathbb{D}) \oplus A_2^1(\mathbb{D}^*)$,
$$\mathbf{K} = \begin{pmatrix} K_1 & K_2 \\ K_3 & K_4 \end{pmatrix},$$
and the vectors
$$\mathbf{u} = \begin{pmatrix} u_1 \\ u_2 \end{pmatrix}, \quad \mathbf{v} = \begin{pmatrix} v_1 \\ v_2 \end{pmatrix} \in A_2^1(\mathbb{D}) \oplus A_2^1(\mathbb{D}^*),$$
where $u_1 = \mathcal{A}(\imath \circ f \circ \imath) \circ \imath \imath'$, $u_2 = -\mathcal{A}(\imath \circ g \circ \imath) \circ \imath \imath'$ and $v_1 = \mathcal{A}(f)$, $v_2 = -\mathcal{A}(g)$. Applying Lemma 3.6 to γ and γ^{-1} and using generalized Grunsky equality, we can succinctly rewrite the two identities as a single equation
$$\mathbf{K}\bar{\mathbf{u}} = -\mathbf{v}.$$
Indeed, Lemma 3.6 applied to γ and γ^{-1} gives
$$K_3 \bar{u}_1 + K_4 \bar{u}_2 = -v_2 \quad \text{and} \quad K_1 \bar{v}_1 + K_2 \bar{v}_2 = -u_1,$$
and from generalized Grunsky equality it follows that the functions
$$w_1(z) = \left(\log \frac{f(z)}{z}\right)' = -\sum_{n=1}^\infty n b_{-n,0} z^{n-1}$$
and
$$w_2(z) = -\left(\log \frac{g(z)}{z}\right)' = -\sum_{n=1}^\infty n b_{n,0} z^{-n-1}$$
satisfy the equations
$$K_1 \bar{w}_1 + K_2 \bar{w}_2 = w_1 \quad \text{and} \quad K_3 \bar{w}_1 + K_4 \bar{w}_2 = w_2.$$
Since $u_1 = v_1 - 2w_1$ and $u_2 = v_2 - 2w_2$, we get the equation $\mathbf{K}\bar{\mathbf{u}} = -\mathbf{v}$. Similarly, we get the equation $\mathbf{K}\bar{\mathbf{v}} = -\mathbf{u}$.

REMARK 3.11. For C^3 curves the result of Theorem 3.5 was obtained by Schiffer and Hawley in [**SH62**]. They have used a completely different approach which can not be generalized to quasi-circles for $T_0(1)$.

The equality $\mathsf{S}_2 = -\frac{1}{12\pi}\mathsf{S}_1$ can be also interpreted as a surgery type formula for determinants of elliptic operators (see [**BFK92, HZ99**]). Namely, let Δ_φ be the Laplace operator of the conformal metric $e^{2\varphi(z)}|dz|^2$ on $\mathbb{D}$ with Dirichlet boundary condition. Its zeta-function regularized determinant $\det \Delta_\varphi$ is given by the Polyakov-Alvarez formula
$$(3.3) \qquad \log \det \Delta_\varphi = -\frac{1}{3\pi} \iint_\mathbb{D} |\varphi_z|^2 d^2z - \frac{1}{6\pi} \oint_{S^1} \varphi(e^{i\theta}) d\theta + \log \det \Delta_0.$$

Now let $\gamma = g^{-1} \circ f \in T_0(1)$ and set, as before, $\tilde{g} = \imath \circ g \circ \imath$, $\tilde{f} = \imath \circ f \circ \imath$. The metric $|\tilde{g}'(z)|^2 |dz|^2$ is a pull-back of the Euclidean metric $|dw|^2$ on $\tilde{\Omega} = \tilde{g}(\mathbb{D})$ by the conformal mapping $\tilde{g}$. Assume that $\phi(z) = \frac{1}{2}\log|\tilde{g}'(z)|^2$ is of C^1 class on S^1, and denote by $\Delta_{\tilde{\Omega}}$ the Laplace operator of the Euclidean metric on $\tilde{\Omega}$ with Dirichlet boundary condition. From (3.3) we immediately get

$$\log \det \Delta_{\tilde{\Omega}} = -\frac{1}{12\pi} \iint_{\mathbb{D}} |\mathcal{A}(\tilde{g})|^2 d^2 z - \frac{1}{3}\log|\tilde{g}'(0)| + \log \det \Delta_{\mathbb{D}}.$$

Now consider the metric $|\tilde{f}'(\frac{1}{z})|^2 |dz|^2$ on $\mathbb{D}$ — a pull-back of the flat metric

$$ds^2 = \frac{|dw|^2}{|\tilde{f}^{-1}(w)|^4}$$

on $\tilde{\Omega}^* = \tilde{f}(\mathbb{D}^*)$ by the conformal mapping $\tilde{f} \circ \imath$. Denoting by $\Delta_{\tilde{\Omega}^*}$ the Laplace operator of the metric ds^2 on $\tilde{\Omega}^*$ with Dirichlet boundary condition, we get from (3.3),

$$\log \det \Delta_{\tilde{\Omega}^*} = -\frac{1}{12\pi} \iint_{\mathbb{D}^*} |\mathcal{A}(\tilde{f})|^2 d^2 z + \log \det \Delta_{\mathbb{D}^*},$$

where we again assumed that $\varphi(z) = \frac{1}{2}\log|\tilde{f}'(\frac{1}{z})|^2$ is of C^1 class on S^1. Here $\Delta_{\mathbb{D}^*}$ is the Laplace operator of the metric $\frac{|dw|^2}{|w|^4}$ on $\mathbb{D}^*$. Note that the metric ds^2 is regular at ∞, so that $\Delta_{\tilde{\Omega}^*}$ is an elliptic operator (cf. [**HZ99**]). The following result now follows from Theorem 3.8 and the symmetry property $\mathsf{S}_1([\mu]) = \mathsf{S}_1([\mu^{-1}])$.

COROLLARY 3.12. *Let $\gamma = g^{-1} \circ f \in T_0(1)$ be of C^3 class on S^1. Then for $\mathcal{C} = f(S^1)$,*

$$\mathrm{Det}_F(\mathcal{C}) = \frac{\det \Delta_{\tilde{\Omega}} \det \Delta_{\tilde{\Omega}^*}}{\det \Delta_{\mathbb{D}} \det \Delta_{\mathbb{D}^*}}.$$

REMARK 3.13. The statement of Corollary 3.12 can be interpreted as a surgery type formula in the spirit of [**BFK92**] for the Laplace operator of a conformal metric on the Riemann sphere $\mathbb{P}^1$, which is the Euclidean metric on the interior domain $\tilde{\Omega} = \imath(\Omega^*)$ and is the metric $ds^2 = \frac{|dw|^2}{|\tilde{f}^{-1}(w)|^4}$ on the exterior domain $\tilde{\Omega}^* = \imath(\Omega)$ (and thus is continuous on $\mathbb{P}^1$). The Fredholm determinant $\mathrm{Det}_F(\mathcal{C})$ is the inverse of the determinant of the Neumann jump operator which corresponds to cutting of $\mathbb{P}^1$ along the contour $\mathcal{C}$ and considering Dirichlet boundary conditions for interior and exterior Laplace operators (cf. [**HZ99**]).

4. Weil-Petersson potential

4.1. Weil-Petersson potential on $T_0(1)$. As in the case of finite dimensional Teichmüller spaces [**TT03**], it follows from the results of the previous section that the function S_1 is a potential for the Weil-Petersson metric on $T_0(1)$. For the convenience of the reader, here we give the details.

THEOREM 4.1. *In terms of the Bers coordinates on the chart V_κ at $\kappa \in T_0(1)$,*

$$\frac{\partial^2 \mathsf{S}_1}{\partial \varepsilon_\mu \bar{\varepsilon}_\nu}([\kappa]) = \iint_{\mathbb{D}^*} \mu(z)\overline{\nu(z)}\rho(z)d^2 z.$$

PROOF. We have

$$\frac{\partial^2 \mathsf{S}_1}{\partial \varepsilon_\mu \bar{\varepsilon}_\nu}([\kappa]) = \frac{\partial}{\partial \bar{\varepsilon}}\bigg|_{\varepsilon=0} \frac{\partial \mathsf{S}_1}{\partial \varepsilon_\mu}([\varepsilon\nu * \kappa]).$$

Using Theorem 3.8 and the fact that at the point $\varepsilon\nu * \kappa \in T_0(1)$ the vector field $\frac{\partial}{\partial \varepsilon_\mu}$ on the chart V_κ is represented by $P(R(\mu, \varepsilon\nu)) \in H^{-1,1}(\mathbb{D}^*)$ on the chart $V_{\varepsilon\nu*\kappa}$ (see Section 2.3 in Chapter 1), we get

$$\frac{\partial \mathsf{S}_1}{\partial \varepsilon_\mu}(\varepsilon\nu * \kappa) = 2 \iint_{\mathbb{D}^*} \mathcal{S}(g_{\varepsilon\nu*\kappa}) P(R(\mu, \varepsilon\nu)) d^2 z$$

$$= 2 \iint_{\mathbb{D}^*} \left(\mathcal{S}(g_\varepsilon) \circ w_{\varepsilon\nu}(w_{\varepsilon\nu})_z^2\right) Q(R(\mu, \varepsilon\nu))(1 - |\varepsilon\nu|^2) d^2 z,$$

where $g_\varepsilon^{-1} \circ f^\varepsilon = w_{\varepsilon\nu} \circ w_\kappa$, $v_\varepsilon = f^\varepsilon \circ f^{-1}$, and $Q(R(\mu, \varepsilon\nu))$ was defined in Section 7.1 in Chapter 1. Since $g_\varepsilon \circ w_{\varepsilon\nu} = v_\varepsilon \circ g$, we have

(4.1) $$\mathcal{S}(g_\varepsilon) \circ w_{\varepsilon\nu}(w_{\varepsilon\nu})_z^2 + \mathcal{S}(w_{\varepsilon\nu}) = \mathcal{S}(v_\varepsilon) \circ g(g')^2 + \mathcal{S}(g),$$

and it follows from the standard variational formula that

$$\frac{\partial}{\partial \bar\varepsilon}\bigg|_{\varepsilon=0} \mathcal{S}(g_\varepsilon) \circ w_{\varepsilon\nu}(w_{\varepsilon\nu})_z^2 = \frac{6}{\pi} \iint_{\mathbb{D}^*} \frac{\overline{\nu(w)}}{(1 - \bar w z)^4} d^2 w.$$

Since according to Theorem 7.4 in Chapter 1

$$\frac{\partial}{\partial \bar\varepsilon}\bigg|_{\varepsilon=0} Q(R(\mu, \varepsilon\nu))$$

is an infinitesimally trivial Beltrami differential, we have

$$\frac{\partial^2}{\partial \varepsilon_\mu \bar\varepsilon_\nu} \mathsf{S}_1([\kappa]) = \frac{12}{\pi} \iint_{\mathbb{D}^*} \iint_{\mathbb{D}^*} \frac{\mu(z)\overline{\nu(w)}}{(1 - \bar w z)^4} d^2 w d^2 z = \iint_{\mathbb{D}^*} \mu(z)\overline{\nu(z)}\rho(z) d^2 z.$$

$\square$

COROLLARY 4.2. *On $T_0(1)$,*

$$\partial\bar\partial \mathsf{S}_1 = -2i\omega_{WP},$$

where ω_{WP} is the symplectic form of the Weil-Petersson metric. In other words, S_1 is a potential of the Weil-Petersson metric on $T_0(1)$.

REMARK 4.3. It follows from Corollary 2.16 and Theorem 3.8 that on $T_0(1)$,

$$\partial\bar\partial \log \det N_\Omega = \frac{i}{6\pi} \omega_{WP}.$$

In the spirit of the last remark in Section 8 of Chapter 1, this result should be compared to the local index theorem for families of $\bar\partial$-operators on compact Riemann surfaces,

$$\partial\bar\partial \log \det \Delta_0 - \partial\bar\partial \log \det N_1 = -\frac{i}{6\pi} \omega_{WP},$$

where N_1 is the period matrix of 1-forms on a compact Riemann surface X and Δ_0 is the Laplace operator of the hyperbolic metric on X (see, e.g., [**ZT87**]).

REMARK 4.4. It follows from Corollary 3.9 that on $T_0(1)$,

$$\partial\vartheta = 0.$$

Here is a direct proof of this result, following our work [**TT03**]. From equation (4.1), we have at $[\kappa] \in T_0(1)$,

$$(L_\nu \vartheta)(z) = \frac{\partial}{\partial \varepsilon_\mu}\bigg|_{\varepsilon=0} \mathcal{S}(g_\varepsilon) \circ w_{\varepsilon\nu}(w_{\varepsilon\nu})_z^2(z)$$

$$= -\frac{12}{\pi} \iint_{\mathbb{D}^*} \nu(w) \left(\frac{g'(w)^2 g'(z)^2}{(g(w)-g(z))^4} - \frac{1}{(w-z)^4} \right) d^2w$$

$$= -\frac{12}{\pi} \iint_{\mathbb{D}^*} \nu(w) \frac{g'(w)^2 g'(z)^2}{(g(w)-g(z))^4} d^2w.$$

Hence,

$$\partial \vartheta(\mu, \nu) = L_\mu \vartheta(\nu) - L_\nu \vartheta(\mu)$$

$$= -\frac{12}{\pi} \iint_{\mathbb{D}^*} \iint_{\mathbb{D}^*} \mu(w) \frac{g'(w)^2 g'(z)^2}{(g(w)-g(z))^4} \nu(z) d^2w d^2z$$

$$+ \frac{12}{\pi} \iint_{\mathbb{D}^*} \iint_{\mathbb{D}^*} \nu(w) \frac{g'(w)^2 g'(z)^2}{(g(w)-g(z))^4} \mu(z) d^2w d^2z$$

$$= 0.$$

4.2. Weil-Petersson potential on $T(1)$. The 1-form ϑ does not naturally extend to the whole Hilbert manifold $T(1)$ (since $\vartheta|_{[\mu]} \in A_2(\mathbb{D}^*)$ if and only if $[\mu] \in T_0(1)$). From Theorem 1.12 we also see that $T_0(1)$ is the maximal subset of $T(1)$ on which the function S_1 is well-defined. However, it is easy to construct a Weil-Petersson potential on $T(1)$ by using right translations. Namely, we index the components of the Hilbert manifold $T(1)$ by the set I (uncountable) and for every $\alpha \in I$ choose $[\mu_\alpha] \in T_\alpha(1) = R_{\mu_\alpha} T_0(1)$ such that $\mu_0 = 0$ for the component $T_0(1)$. This represents $T(1)$ as a disjoint union

$$T(1) = \bigsqcup_{\alpha \in I} T_\alpha(1).$$

Define

$$\mathsf{S}([\nu]) = \mathsf{S}_1([\nu * \mu_\alpha^{-1}]) \quad \text{for} \quad [\nu] \in T_\alpha(1).$$

It follows from the right-invariance of the Weil-Petersson metric that the function S is a Weil-Petersson potential on $T(1)$.

5. The period mapping

The generalization of the classical period mapping to the homogeneous space $\text{Möb}(S^1) \backslash \text{Diff}_+(S^1)$ was outlined by Kirillov and Yuriev in [**KY88**] and developed by Nag [**Nag92**]. In particular, in [**Nag92**] it is explained in what sense this is a generalization of classical period mapping as an association between the complex structures and corresponding spaces of holomorphic 1-forms. Subsequently in [**NS95**], Nag and Sullivan extended the period mapping to the universal Teichmüller space $T(1)$. Here we prove that the Kirillov-Yuriev-Nag-Sullivan (KYNS) period mapping coincides with the mapping $\hat{\mathscr{P}}$ defined in Remark 2.11.

5.1. KYNS period mapping. Following [**NS95**], let $\mathcal{H}$ be the real Hilbert space

$$\mathcal{H} = H^{1/2}(S^1, \mathbb{R})/\mathbb{R}$$
$$= \left\{ f : S^1 \to \mathbb{R} \,\Big|\, f(e^{i\theta}) = {\sum_{n=-\infty}^{\infty}}' c_n e^{in\theta}, \ \sum_{n=1}^{\infty} n|c_n|^2 < \infty \right\},$$

and let Θ be the symplectic form[2] on $\mathcal{H}$:

$$\Theta(f, g) = \frac{1}{2\pi} \oint_{S^1} g\, df.$$

By complex linearity, the symplectic form Θ extends to the complexification of $\mathcal{H}$ — the complex Hilbert space $\mathcal{H}_\mathbb{C}$,

$$\mathcal{H}_\mathbb{C} = H^{1/2}(S^1, \mathbb{C})/\mathbb{C}$$
$$= \left\{ f : S^1 \to \mathbb{C} \,\Big|\, f(e^{i\theta}) = {\sum_{n=-\infty}^{\infty}}' c_n e^{in\theta}, \ {\sum_{n=-\infty}^{\infty}}' |n||c_n|^2 < \infty \right\}.$$

With respect to this symplectic form, the Hilbert space $\mathcal{H}_\mathbb{C}$ has a canonical decomposition into two closed isotropic subspaces

$$\mathcal{H}_\mathbb{C} = W_+ \oplus W_-,$$

where

$$W_+ = \left\{ f : S^1 \to \mathbb{C} \,\Big|\, f(e^{i\theta}) = \sum_{n=1}^{\infty} a_n e^{in\theta}, \ \sum_{n=1}^{\infty} n|a_n|^2 < \infty \right\},$$
$$W_- = \left\{ g : S^1 \to \mathbb{C} \,\Big|\, g(e^{i\theta}) = \sum_{n=1}^{\infty} b_n e^{-in\theta}, \ \sum_{n=1}^{\infty} n|b_n|^2 < \infty \right\}.$$

Let $\mathfrak{D}_\infty$ be the infinite dimensional analog of Siegel disk [**Sie64**],

$$\mathfrak{D}_\infty = \left\{ Z \in \mathscr{B}(W_-, W_+) \,:\, \Theta(Zf, g) = \Theta(Zg, f) \ \text{and} \ I - Z\bar{Z} > 0 \right\}.$$

Here $\mathscr{B}(W_-, W_+)$ is the Banach space of all bounded linear operators from W_- to W_+, and $\bar{Z} = JZJ : W_+ \to W_-$, where J is the standard conjugation operator on $\mathcal{H}_\mathbb{C}$ defined by $JW_+ = W_-$. With respect to the standard bases

$$\left\{ e_n = \tfrac{1}{\sqrt{n}} e^{in\theta} \right\}_{n \in \mathbb{N}} \quad \text{and} \quad \left\{ f_n = \tfrac{1}{\sqrt{n}} e^{-in\theta} \right\}_{n \in \mathbb{N}}$$

of the subspaces W_+ and W_-, an operator $Z \in \mathfrak{D}_\infty$ is represented by an infinite matrix, and the condition $\Theta(Zf, g) = \Theta(Zg, f)$ translates as $Z = Z^t$. Let $\mathrm{Sp}(\mathcal{H})$ be the group of bounded symplectomorphisms on $\mathcal{H}$. Elements of $\mathrm{Sp}(\mathcal{H})$, extended complex linearly to $\mathcal{H}_\mathbb{C}$, in the basis $\{e_n\}_{n \in \mathbb{N}} \cup \{f_n\}_{n \in \mathbb{N}}$ of $\mathcal{H}_\mathbb{C}$ can be represented by matrices

(5.1) $$\begin{pmatrix} A & B \\ \bar{B} & \bar{A} \end{pmatrix}, \quad \text{where} \quad AA^* - BB^* = I, \ AB^t = BA^t.$$

The group $\mathrm{Sp}(\mathcal{H})$ acts transitively on $\mathfrak{D}_\infty$ by

$$Z \mapsto (AZ + B)(\bar{B}Z + \bar{A})^{-1},$$

---

[2] We use a different sign convention since our complex structure has a different sign compared to [**KY88, Nag92, NS95**].

5. THE PERIOD MAPPING

and the stabilizer of the point $Z = 0$ is the unitary subgroup U of $\mathrm{Sp}(\mathcal{H})$ consisting of bounded symplectomorphisms with $B = 0$. Thus the canonical quotient map $Q : \mathrm{Sp}(\mathcal{H}) \to \mathfrak{D}_\infty$,

$$Q\left(\begin{pmatrix} A & B \\ \bar{B} & \bar{A} \end{pmatrix}\right) = (AZ + B)(\bar{B}Z + \bar{A})^{-1}\big|_{Z=0} = B\bar{A}^{-1},$$

induces the isomorphism

$$\mathrm{Sp}(\mathcal{H})/U \simeq \mathfrak{D}_\infty.$$

In [**NS95**], Nag and Sullivan proved that the assignment

$$\mathrm{Homeo}_{qs}(S^1) \ni \gamma \mapsto \hat{\Pi}(\gamma) \in \mathscr{B}(\mathcal{H}_\mathbb{C}),$$

where

$$\hat{\Pi}(\gamma)(f) = f \circ \gamma - \frac{1}{2\pi} \oint_{S^1} f \circ \gamma \, d\theta, \quad f \in \mathcal{H}_\mathbb{C},$$

defines a right action of the group $\mathrm{Homeo}_{qs}(S^1)$ on the Hilbert space $\mathcal{H}_\mathbb{C}$ by symplectomorphisms. Thus the mapping

$$\hat{\Pi} : \mathrm{Homeo}_{qs}(S^1) \to \mathrm{Sp}(\mathcal{H})$$

satisfies $\hat{\Pi}(\gamma_1 \circ \gamma_2) = \hat{\Pi}(\gamma_2)\hat{\Pi}(\gamma_1)$. On the other hand, an operator $\hat{\Pi}(\gamma)$ preserves the subspaces W_+ and W_-, i.e., $\hat{\Pi}(\gamma) \in U$, if and only if $\gamma \in \mathrm{M\ddot{o}b}(S^1)$. The induced mapping

$$\Pi = Q \circ \hat{\Pi} : T(1) = \mathrm{M\ddot{o}b}(S^1) \backslash \mathrm{Homeo}_{qs}(S^1) \to \mathrm{Sp}(\mathcal{H})/U \simeq \mathfrak{D}_\infty$$

is what we call KYNS period mapping of $T(1)$.

With respect to the basis $\{e_n\}_{n \in \mathbb{N}} \cup \{f_n\}_{n \in \mathbb{N}}$ of $\mathcal{H}_\mathbb{C}$, the mapping

$$\hat{\Pi} : \mathrm{Homeo}_{qs}(S^1) \to \mathrm{Sp}(\mathcal{H})$$

is given by the matrix

$$\hat{\Pi}(\gamma) = \begin{pmatrix} \mathfrak{A} & \mathfrak{B} \\ \bar{\mathfrak{B}} & \bar{\mathfrak{A}} \end{pmatrix}, \quad \gamma \in \mathrm{Homeo}_{qs}(S^1),$$

where

$$\mathfrak{A}_{mn}(\gamma) = \frac{1}{2\pi}\sqrt{\frac{m}{n}} \oint_{S^1} (\gamma(e^{i\theta}))^n e^{-im\theta} d\theta,$$

$$\mathfrak{B}_{mn}(\gamma) = \frac{1}{2\pi}\sqrt{\frac{m}{n}} \oint_{S^1} (\gamma(e^{i\theta}))^{-n} e^{-im\theta} d\theta.$$

As a result, the KYNS period matrix $\Pi : T(1) \to \mathfrak{D}_\infty$ is given by the matrix

$$\Pi([\mu]) = \mathfrak{B}\bar{\mathfrak{A}}^{-1}, \quad [\mu] \in T(1),$$

where $\mathfrak{A} = \mathfrak{A}(w_\mu)$ and $\mathfrak{B} = \mathfrak{B}(w_\mu)$. On the other hand, it follows from (5.1) that

$$\hat{\Pi}(\gamma^{-1}) = \hat{\Pi}(\gamma)^{-1} = \begin{pmatrix} \mathfrak{A}(\gamma)^* & -\mathfrak{B}(\gamma)^t \\ -\mathfrak{B}(\gamma)^* & \mathfrak{A}(\gamma)^t \end{pmatrix}.$$

PROPOSITION 5.1. *The Grunsky matrices B_l, $l = 1, 2, 3, 4$, corresponding to $\gamma \in S^1 \backslash \mathrm{Homeo}_{qs}(S^1)$, and the elements of the matrix $\hat{\Pi}(\gamma)$ are related by*

$$B_1 = \mathfrak{B}\bar{\mathfrak{A}}^{-1}, \quad B_2 = (\mathfrak{A}^*)^{-1},$$

$$B_3 = \bar{\mathfrak{A}}^{-1}, \quad B_4 = -\mathfrak{B}^*(\mathfrak{A}^*)^{-1}.$$

PROOF. Let $\gamma = g^{-1} \circ f$ be the conformal welding of $\gamma \in S^1 \backslash \text{Homeo}_{qs}(S^1)$, and let P_n and Q_n be the Faber polynomials associated to the pair (f,g). Denoting by $\text{P}_+ : \mathcal{H}_{\mathbb{C}} \to W_+$ and $\text{P}_- : \mathcal{H}_{\mathbb{C}} \to W_-$ the orthogonal projection operators, we get

$$\mathfrak{A}^* = \text{P}_+ \hat{\Pi}(\gamma^{-1}) \Big|_{W_+}.$$

By definition of Faber polynomials, $(P_n^0 \circ f)\big|_{S^1} \in W_+$, where $P_n^0(z) = P_n(z) - P_n(0)$ and $n \geq 1$. We have on S^1,

$$\text{P}_+ \hat{\Pi}(\gamma^{-1})(P_n^0 \circ f) = \text{P}_+ \left((P_n^0 \circ f) \circ f^{-1} \circ g\right) = \text{P}_+(P_n^0 \circ g).$$

Since $\text{P}_+(P_n^0 \circ g)(e^{i\theta}) = e^{in\theta}$ and $(P_n^0 \circ f)(e^{i\theta}) = n \sum_{m=1}^{\infty} b_{-m,n} e^{im\theta}$, we obtain

$$\sum_{k=1}^{\infty} \mathfrak{A}^*_{mk}(B_2)_{kn} = \delta_{mn},$$

i.e. $\mathfrak{A}^* B_2 = \text{Id}$. Similarly, let $Q_n^0 = Q_n - Q_n(\infty)$. By definition of Faber polynomials, $(Q_n^0 \circ f)(e^{i\theta}) - e^{-in\theta} \in W_+$ for $n \geq 1$. We have on S^1,

$$\text{P}_+ \hat{\Pi}(\gamma^{-1})((Q_n^0(f(e^{i\theta})) - e^{-in\theta}) = \text{P}_+ \left((Q_n^0 \circ g) - (\gamma^{-1})^{-n}\right) = -\text{P}_+((\gamma^{-1})^{-n}).$$

Since $-\mathfrak{B}(\gamma^{-1}) = \mathfrak{B}^t$, we have $-\text{P}_+((\gamma^{-1})^{-n})(e^{i\theta}) = \sum_{k=1}^{\infty} \sqrt{\frac{n}{k}} \mathfrak{B}_{nk} e^{ik\theta}$. Using $Q_n^0(f(e^{i\theta})) - e^{-in\theta} = n \sum_{m=1}^{\infty} b_{-m,-n} e^{im\theta}$, we obtain

$$\sum_{k=1}^{\infty} \mathfrak{A}^*_{mk}(B_1)_{kn} = \mathfrak{B}_{nm},$$

i.e., $\mathfrak{A}^* B_1 = \mathfrak{B}^t$, which is equivalent to $B_1 = B_1^t = \mathfrak{B}\bar{\mathfrak{A}}^{-1}$. Using $B_3 = B_2^t$ and $B_4(\gamma) = \overline{B_1(\gamma^{-1})}$ concludes the proof. $\square$

COROLLARY 5.2. *The KYNS period mapping Π coincides with our period mapping $\hat{\mathscr{P}}$ defined in Remark 2.11.*

PROOF. Due to Proposition 5.1, $B_1 = \mathfrak{B}\bar{\mathfrak{A}}^{-1}$. $\square$

REMARK 5.3. In [**NS95**] it was stated that the period mapping $\Pi : T(1) \to \mathfrak{D}_{\infty}$ is a holomorphic mapping of Banach manifolds. However, it was only shown that the induced mapping $D\Pi$ of tangent spaces is complex linear injection, which is not enough to claim holomorphy for infinite dimensional manifolds. In Appendix B we prove that the mapping $\hat{\mathscr{P}} : T(1) \to \mathscr{B}(\ell^2)$ is holomorphic, which completes the proof in [**NS95**].

Following G. Segal [**Seg81**], we introduce the subgroup $\text{Sp}_0(\mathcal{H})$ of the symplectic group $\text{Sp}(\mathcal{H})$ for which $B \in \mathscr{S}_2(W_-, W_+)$ — the Hilbert space of Hilbert-Schmidt operators from W_- to W_+. The group $\text{Sp}_0(\mathcal{H})$ acts transitively on the restricted Siegel disc

$$\mathfrak{D}_{\infty}^0 = \mathfrak{D}_{\infty} \cap \mathscr{S}_2(W_-, W_+).$$

Corollaries 5.2 and 2.9 immediately imply the following result.

COROLLARY 5.4. *For $[\mu] \in T(1)$, $\Pi([\mu]) \in \mathfrak{D}_{\infty}^0$ if and only if $[\mu] \in T_0(1)$.*

REMARK 5.5. In view of the above corollary, define the restricted period mapping
$$\Pi_0 = \Pi|_{T_0(1)} : T_0(1) \to \mathfrak{D}_\infty^0.$$
Since by Corollary 5.2 $\Pi_0 = \mathscr{P}$, by Theorem 2.10 Π_0 is a holomorphic mapping of Hilbert manifolds. The homogenuous space $\mathfrak{D}_\infty^0$ carries a natural $\mathrm{Sp}_0(\mathcal{H})$-invariant Kähler metric with the Kähler potential $\Phi(Z) = \log \mathrm{Det}(1-Z\bar{Z})$. It was first shown by Kirillov and Yuriev [**KY88**] and later by Nag [**Nag92**] that the pullback of this metric to $\mathrm{M\ddot{o}b}(S^1)\backslash \mathrm{Diff}_+(S^1)$ by the period mapping coincides, up to a constant, with the Weil-Petersson metric. It immediately follows from Corollary 5.2 that
$$\mathsf{S}_2 = \log \mathrm{Det}(I - Z\bar{Z}),$$
so that the pullback of the natural Kähler metric on $\mathfrak{D}_\infty^0$ by the restricted period mapping to $T_0(1)$ coincides, up to a constant, with the Weil-Petersson metric on $T_0(1)$. Thus we have established the relations between all natural potential functions on $T_0(1)$: up to a constant factor, they are indeed all equal!

5.2. Embeddings into the Segal-Wilson universal Grassmannian. Let $\mathscr{V}$ be an infinite-dimensional separable complex Hilbert space and let
$$\mathscr{V} = V_+ \oplus V_-$$
be its decomposition into the direct sum of infinite-dimensional closed subspaces V_+ and V_-. The Segal-Wilson universal Grassmannian $\mathrm{Gr}(\mathscr{V})$ [**SW85**, **PS86**] is defined as a set of closed subspaces W of $\mathscr{V}$ satisfying the following conditions.

UG1. The orthogonal projection $\mathrm{pr}_+ : W \to V_+$ is a Fredholm operator.

UG2. The orthogonal projection $\mathrm{pr}_- : W \to V_-$ is a Hilbert-Schmidt operator.

Equivalently, $W \in \mathrm{Gr}(\mathscr{V})$, if W is the image of an operator $\mathrm{w} : V_+ \to W$ such that $\mathrm{pr}_+\mathrm{w}$ is Fredholm and $\mathrm{pr}_-\mathrm{w}$ is Hilbert-Schmidt. The Segal-Wilson Grassmannian $\mathrm{Gr}(\mathscr{V})$ is a Hilbert manifold modeled on the Hilbert space $\mathscr{S}_2(V_+, V_-)$ of Hilbert-Schmidt operators from V_+ to V_-.

For our purposes, let $\mathscr{V} = \mathcal{H}_\mathbb{C}$ and $V_+ = W_-$, $V_- = W_+$. To every $[\mu] \in T_0(1)$ we associate a closed subspace $W_\mu \subset \mathcal{H}_\mathbb{C}$ spanned by the functions $w_n(e^{i\theta}) = \frac{1}{\sqrt{n}} Q_n(\mathrm{f}^\mu(e^{i\theta}))$, where $w_\mu = \mathrm{g}_\mu^{-1} \circ \mathrm{f}^\mu$ is the corresponding conformal welding. Explicitly, in terms of the basis $\{e_n\}_{n\in\mathbb{N}} \cup \{f_n\}_{n\in\mathbb{N}}$ of $\mathcal{H}_\mathbb{C}$,
$$w_n = f_n + \sum_{m=1}^\infty \sqrt{nm}\, b_{-n,-m} e_m,\ n \in \mathbb{N}.$$
We have $W_\mu = \mathrm{w}(V_+)$, where $\mathrm{w}(f_n) = w_n$, $n \in \mathbb{N}$. Thus the mapping $\mathrm{pr}_+\mathrm{w} = I$ — the identity operator on V_+, is obviously Fredholm, and the mapping $\mathrm{pr}_-\mathrm{w} = B_1(\mathrm{f}^\mu)$ is Hilbert-Schmidt since $[\mu] \in T_0(1)$. According to Theorem 2.10, the mapping
$$\mathscr{E} : T_0(1) \to \mathrm{Gr}(\mathcal{H}_\mathbb{C})$$
given by $\mathscr{E}([\mu]) = W_\mu$ is a holomorphic inclusion of $T_0(1)$ into the Segal-Wilson universal Grassmannian. For the homogeneous space $\mathrm{M\ddot{o}b}(S^1)\backslash \mathrm{Diff}_+(S^1)$ this mapping was first considered in [**KY88**].

REMARK 5.6. Seemingly another mapping of $\mathrm{M\ddot{o}b}(S^1)\backslash \mathrm{Diff}_+(S^1)$ into the Segal-Wilson Grassmannian was considered in [**STZ99**]. Namely, extend $\mu \in L^\infty(\mathbb{D}^*)$ by zero to $\mathbb{D}$ and let V_μ be the space of distributional solutions of the Beltrami equation $w_{\bar{z}} = \mu w_z$ on $\mathbb{C}$ having a single pole at 0. The mapping in [**STZ99**] was defined by the assignment $[\mu] \to V_\mu|_{S^1}$. It is easy to see that the space V_μ is spanned by the functions $w_n(z) = Q_n(f^\mu)(z)$, $n \in \mathbb{N}$, so that $V_\mu|_{S^1} = W_\mu$ and the mapping in [**STZ99**] coincides with the Kirillov-Yuriev mapping [**KY88**].

REMARK 5.7. The inclusion $\mathscr{E} : T_0(1) \to \mathrm{Gr}(\mathcal{H}_\mathbb{C})$ is a holomorphic mapping due to the holomorphic dependence of f^μ on μ. Since f^μ is holomorphic on $\mathbb{D}$, the subspaces W_μ correspond to the different uniformizations of the same Riemann surface $\Omega = f^\mu(\mathbb{D}) \simeq \mathbb{D}$. However, one can consider another mapping where the associated subspaces in the universal Grassmannian correspond to Riemann surfaces of different complex structure. Namely, set, as before, $\mathscr{V} = \mathcal{H}_\mathbb{C}$ and let $V_+ = W_+$ and $V_- = W_-$. We denote the corresponding Segal-Wilson Grassmannian by $\widetilde{\mathrm{Gr}}(\mathcal{H}_\mathbb{C})$, and define the mapping

$$\tilde{\mathscr{E}} : T_0(1) \to \widetilde{\mathrm{Gr}}(\mathcal{H}_\mathbb{C})$$

by assigning to every point $[\mu] \in T_0(1)$ the closed subspace $\tilde{W}_\mu \subset \mathcal{H}_\mathbb{C}$ spanned by the functions $\tilde{w}_n(e^{i\theta}) = \frac{1}{\sqrt{n}} P_n(g_\mu(e^{i\theta}))$, $n \in \mathbb{N}$. We have $\tilde{W}_\mu = \tilde{w}(V_+)$, where $\tilde{w}(e_n) = \tilde{w}_n$, $n \in \mathbb{N}$, and $\mathrm{pr}_+ \tilde{w} = I$ — the identity operator on V_+ and $\mathrm{pr}_- \tilde{w} = B_4(g_\mu)$ is Hilbert-Schmidt since $[\mu] \in T_0(1)$. The mapping $\tilde{\mathscr{E}}$ is not holomorphic. However, since $JW_+ = W_-$, where J is the standard conjugation operator on $\mathcal{H}_\mathbb{C}$, we have $\widetilde{\mathrm{Gr}}(\mathcal{H}_\mathbb{C}) = J(\mathrm{Gr}(\mathcal{H}_\mathbb{C}))$, so that $\widetilde{\mathrm{Gr}}(\mathcal{H}_\mathbb{C})$ is a mirror image of $\mathrm{Gr}(\mathcal{H}_\mathbb{C})$. Denoting by $\mathcal{I}$ the inversion on the topological group $T_0(1)$, we get

$$\tilde{\mathscr{E}} = J \circ \mathscr{E} \circ \mathcal{I}.$$

REMARK 5.8. One can describe the "Schottky locus", i.e., the image $\mathscr{E}(T_0(1))$ in the Segal-Wilson Grassmannian $\mathrm{Gr}(\mathcal{H}_\mathbb{C})$. Indeed, since the corresponding points in $\mathrm{Gr}(\mathcal{H}_\mathbb{C})$ are associated with the Grunsky operators B_1, it is equivalent to the characterization of the image of the restricted period map $\Pi_0 : T_0(1) \to \mathfrak{D}_\infty^0$. Let $C = \{C_{mn}\}_{m,n \in \mathbb{N}} \in \mathfrak{D}_\infty^0$, which we realized as symmetric, Hilbert-Schmidt operator on ℓ^2 satisfying $I - C\bar{C} > 0$. Then $C \in \Pi_0(T_0(1))$ if and only if the the following conditions are satisfied.

S1.
$$1 + \sum_{m=1}^{\infty} \frac{C_{m1}}{\sqrt{m}} \frac{z_1^m - z_2^m}{z_1^{-1} - z_2^{-1}} = \exp\left(-\sum_{m,n=1}^{\infty} \frac{C_{mn}}{\sqrt{mn}} z_1^m z_2^n\right).$$

S2. There exist $D = \{D_{mn}\}_{m,n \in \mathbb{N}} \in \mathfrak{D}_\infty^0$ and $B \in \mathscr{B}(\ell^2)$ such that

$$I - C\bar{C} = BB^* \quad \text{and} \quad I - D\bar{D} = \bar{B}^* B.$$

S3.
$$1 + \sum_{m=1}^{\infty} \frac{D_{m1}}{\sqrt{m}} \frac{z_1^{-m} - z_2^{-m}}{z_1 - z_2} = \exp\left(-\sum_{m,n=1}^{\infty} \frac{D_{mn}}{\sqrt{mn}} z_1^{-m} z_2^{-n}\right).$$

Equations **S1** and **S3** are understood as infinite sequence of relations between elements of the matrices C and D obtained by comparing coefficients of $z_1^m z_2^n$ and $z_1^{-m} z_2^{-n}$ respectively. Equations **S1** and **S3** are nothing but dispersionless Hirota equations (see, e.g., [**Teo03**]). They are just a reformulation of the definition of the Grunsky coefficients of the univalent functions f and g and the identities

$$\frac{1}{f(z)} + c = Q_1(f(z)) = \frac{1}{z} + \sum_{m=1}^{\infty} b_{-1,-m} z^m,$$

$$\frac{g(z)}{b} + d = P_1(g(z)) = z + \sum_{m=1}^{\infty} b_{1m} z^{-m},$$

where c and d are constants. See [**Teo03**] for details.

APPENDIX A

The Hilbert Manifold Structure of $\mathcal{T}_0(1)$

Here we show that the Hilbert manifold $\mathcal{T}_0(1)$, modeled on the Hilbert space $A_2(\mathbb{D}) \oplus \mathbb{C}$, can also be modeled on the Hilbert space $A_2^1(\mathbb{D})$, which induce the same Hilbert manifold structure. This result is parallel to the one in the Appendix of [**Teo04**].

Let β be the Bers embedding $\mathcal{T}(1) \hookrightarrow A_\infty(\mathbb{D}) \oplus \mathbb{C}$,

$$\mathcal{T}(1) \ni \gamma = g^{-1} \circ f \mapsto \left(\mathcal{S}(f), \tfrac{1}{2}\mathcal{A}(f)(0)\right),$$

and let $\hat{\beta} : \mathcal{T}(1) \to A_\infty^1(\mathbb{D})$ be the pre-Bers embedding of $\mathcal{T}(1)$ into $A_\infty^1(\mathbb{D})$,

$$\mathcal{T}(1) \ni \gamma = g^{-1} \circ f \mapsto \mathcal{A}(f).$$

By Theorem 1.12, $\hat{\beta}(\gamma) \in A_2^1(\mathbb{D})$ if and only if $\gamma \in \mathcal{T}_0(1)$.

LEMMA A.1. *The map* $\widehat{\Psi} : A_2^1(\mathbb{D}) \to A_2(\mathbb{D}) \oplus \mathbb{C}$,

$$\widehat{\Psi}(\psi) = \left(\Psi(\psi), \tfrac{1}{2}\psi(0)\right),$$

where $\Psi(\psi) = \psi_z - \tfrac{1}{2}\psi^2$, *is a one to one holomorphic mapping on Hilbert spaces.*

PROOF. Firstly, the map $\Psi : A_2^1(\mathbb{D}) \to A_2(\mathbb{D})$ is holomorphic. That is, for every $\psi, \varphi \in A_2^1(\mathbb{D})$, the map $\mathbb{C} \ni t \mapsto \Psi(\psi + t\varphi)$ is holomorphic in a neighbourhood of $0 \in \mathbb{C}$. Indeed,

$$\left\| \frac{\Psi(\psi + t\varphi) - \Psi(\psi + t_0\varphi)}{t - t_0} - \frac{d}{dt}\bigg|_{t=t_0} \Psi(\psi + t\varphi) \right\|_{A_2(\mathbb{D})}$$

$$= \frac{|t - t_0|}{2} \|\varphi^2\|_{A_2(\mathbb{D})} \leq \frac{|t - t_0|}{4} \|\varphi\|_{A_\infty^1(\mathbb{D})} \|\varphi\|_{A_2^1(\mathbb{D})} = O(|t - t_0|).$$

Secondly, by Lemma 1.3,

$$|\psi(0)| \leq \sqrt{\frac{1}{\pi}} \|\psi\|_{A_2^1(\mathbb{D})},$$

so that $\psi \mapsto \tfrac{1}{2}\psi(0)$ is a bounded complex-linear map. The injectivity of $\widehat{\Psi}$ has been proved in the Appendix of [**Teo04**]. □

COROLLARY A.2. *The set* $\hat{\beta}(\mathcal{T}_0(1)) \subset A_2^1(\mathbb{D})$ *is open in* $A_2^1(\mathbb{D})$.

PROOF. It readily follows from the results in Section 3.3 of Part I that $\beta(\mathcal{T}_0(1))$ is open in $A_2(\mathbb{D}) \oplus \mathbb{C}$. The assertion now follows from the lemma above since $\hat{\beta}(\mathcal{T}_0(1)) = \widehat{\Psi}^{-1}(\beta(\mathcal{T}_0(1)))$. □

THEOREM A.3. *The embeddings* $\beta : \mathcal{T}_0(1) \hookrightarrow A_2(\mathbb{D}) \oplus \mathbb{C}$ *and* $\hat{\beta} : \mathcal{T}_0(1) \hookrightarrow A_2^1(\mathbb{D})$ *induce the same Hilbert manifold structure on* $\mathcal{T}_0(1)$.

PROOF. The map $\widehat{\Psi} : \hat{\beta}(\mathcal{T}_0(1)) \to \beta(\mathcal{T}_0(1))$ is a holomorphic bijection between complex manifolds. To show that $\widehat{\Psi}$ is biholomorphic, by inverse function theorem (see, e.g., [**Lan95**]) we need to prove that for every $\psi \in \hat{\beta}(\mathcal{T}_0(1))$ the linear map $D_\psi \widehat{\Psi}$ is a topological isomorphism between the Hilbert spaces $A_2^1(\mathbb{D})$ and $A_2(\mathbb{D}) \oplus \mathbb{C}$. Let $\gamma = g^{-1} \circ f \in \mathcal{T}_0(1)$ and $\psi = \mathcal{A}(f) \in \hat{\beta}(\mathcal{T}_0(1))$. The linear map $D_\psi \widehat{\Psi} : A_2^1(\mathbb{D}) \to A_2(\mathbb{D}) \oplus \mathbb{C}$ is given by

$$\varphi \mapsto \left(\varphi_z - \psi\varphi, \tfrac{1}{2}\varphi(0)\right).$$

For every $(\phi, c) \in A_2(\mathbb{D}) \oplus \mathbb{C}$, the holomorphic function φ on $\mathbb{D}$, defined by

$$\varphi(z) = f'(z) \left(\int_0^z \frac{\phi(u)}{f'(u)} du + 2c \right),$$

satisfies

$$\varphi_z - \psi\varphi = \phi \quad \text{and} \quad \tfrac{1}{2}\varphi(0) = c.$$

We claim that $\varphi \in A_2^1(\mathbb{D})$, so that the map $D_\psi \widehat{\Psi}$ is onto. Indeed, repeating the proof of Lemma 1.11, we get for $z \in \mathbb{D}$,

$$|\phi(z)|^2 \geq \tfrac{1}{2}|\varphi_z|^2 - |\psi(z)|^2 |\varphi(z)|^2.$$

By Becker-Pommerenke theorem, there exists $r' > 0$ such that

$$|(1 - |z|^2)\psi(z)| \leq \frac{1}{2\sqrt{2}} \quad \text{for all} \quad r' < |z| < 1.$$

Thus for $r' < |z| < 1$,

$$2(1 - |z|^2)^2 |\phi(z)|^2 \geq (1 - |z|^2)^2 |\varphi_z(z)|^2 - \tfrac{1}{4}|\varphi(z)|^2,$$

and the result follows as in the proof of Lemma 1.11. Uniqueness theorem for differential equations guarantees that the map $D_\psi \widehat{\Psi}$ is one-to-one. Finally, by using the same arguments as in the proof of Lemma 1.5 and Lemma 1.3, there exists $C > 0$ such that for all $\varphi \in A_2^1(\mathbb{D})$,

$$\|D_\psi \widehat{\Psi}(\varphi)\|_{A_2(\mathbb{D})} \leq C \|\varphi\|_{A_2^1(\mathbb{D})}.$$

Hence $D_\psi \widehat{\Psi}$ is a bounded linear bijection between Hilbert spaces. □

COROLLARY A.4. *Let $\{\gamma_n\}_{n=1}^\infty$ be a sequence of points in $\mathcal{T}_0(1)$, $\gamma_n = g_n^{-1} \circ f_n$, and let $\gamma = g^{-1} \circ f \in \mathcal{T}_0(1)$. Then the following conditions are equivalent.*

(i) *In $\mathcal{T}_0(1)$ topology,*

$$\lim_{n \to \infty} \gamma_n = \gamma.$$

(ii) *In $A_2^1(\mathbb{D})$ topology,*

$$\lim_{n \to \infty} \mathcal{A}(f_n)(z) = \mathcal{A}(f)(z).$$

(iii) *In $A_2^1(\mathbb{D}^*)$ topology,*

$$\lim_{n \to \infty} \left(\mathcal{A}(g_n)(z) - 2\frac{g_n'(z)}{g_n(z)} + \frac{2}{z} \right) = \mathcal{A}(g)(z) - 2\frac{g'(z)}{g(z)} + \frac{2}{z}.$$

PROOF. The equivalence (i)⇔(ii) follows from Theorem A.3. Since $\mathcal{T}_0(1)$ is a topological group, $\lim_{n\to\infty} \gamma_n = \gamma$ if and only if $\lim_{n\to\infty} \gamma_n^{-1} = \gamma^{-1}$. Now let $j(z) = \frac{1}{z}$ and let r be the dilation $z \mapsto \overline{g'(\infty)}\, z$. We have $\gamma^{-1} = \tilde{g}^{-1} \circ \tilde{f}$, where $\tilde{f} = r \circ j \circ g \circ j$ and

$$\mathcal{A}(\tilde{f}) = \mathcal{A}(r \circ j \circ g \circ j) = \overline{\left(\mathcal{A}(g) - 2\frac{g'}{g} + 2\bar{j}\right)} \circ jj_{\bar{z}},$$

so that the equivalence (i)⇔(iii) follows from the equivalence (i)⇔(ii). □

Next, consider the mappings $\beta^* : \mathcal{T}_0(1) \to A_2(\mathbb{D}^*)$,

$$\beta^*(\gamma) = \overline{\beta(\gamma^{-1}) \circ jj_{\bar{z}}^2} = \mathcal{S}(g),$$

and $\hat{\beta}^* : \mathcal{T}_0(1) \to A_2^1(\mathbb{D}^*)$,

$$\hat{\beta}^*(\gamma) = \mathcal{A}(g),$$

where $\gamma = g^{-1} \circ f \in \mathcal{T}_0(1)$. Also, consider the mapping $\Psi^* : A_2^1(\mathbb{D}^*) \to A_2(\mathbb{D}^*)$, defined by

$$\Psi^*(\psi) = \psi_z - \tfrac{1}{2}\psi^2,$$

and let

$$\widetilde{A_2^1}(\mathbb{D}^*) = \left\{ \psi \in A_2^1(\mathbb{D}^*) : \psi(z) = O\left(\frac{1}{z^3}\right) \text{ as } z \to \infty \right\}.$$

We have the following result.

LEMMA A.5.
(i) *The map* $\Psi^* : A_2^1(\mathbb{D}^*) \to A_2(\mathbb{D}^*)$ *is a holomorphic mapping on Hilbert spaces and its restriction to the subspace* $\widetilde{A_2^1}(\mathbb{D}^*)$ *is injective.*
(ii) *The set* $\hat{\beta}^*(\mathcal{T}_0(1)) = \hat{\beta}^*(\mathcal{T}_0(1))$ *is open in* $A_2^1(\mathbb{D}^*)$.

PROOF. Holomorphy of Ψ^* is proved along the same lines as Lemma A.1. From the proof of Theorem A.5 in [Teo04] it follows that the restriction of the map Ψ^* to the subspace $\widetilde{A_2^1}(\mathbb{D}^*)$ is one to one. To prove part (ii), observe that $\beta^* = \Psi^* \circ \hat{\beta}^*$ and $\beta^*(\mathcal{T}_0(1)) = \beta^*(\mathcal{T}_0(1))$. Since $\hat{\beta}^*(\mathcal{T}_0(1)), \hat{\beta}^*(\mathcal{T}_0(1)) \in \widetilde{A_2^1}(\mathbb{D}^*)$ and the restriction of Ψ^* to $\widetilde{A_2^1}(\mathbb{D}^*)$ is injective, we have the equality $\hat{\beta}^*(\mathcal{T}_0(1)) = \hat{\beta}^*(\mathcal{T}_0(1))$. The proof that this set is open in $A_2^1(\mathbb{D}^*)$ is analogous to the proof of Corollary A.2. □

COROLLARY A.6. *Let* $\{\gamma_n\}_{n=1}^\infty$, $\gamma_n = g_n^{-1} \circ f_n$, *be a sequence of points in* $\mathcal{T}_0(1)$ *such that*

$$\lim_{n\to\infty} \gamma_n = \gamma = g^{-1} \circ f \in \mathcal{T}_0(1).$$

Then the following statements hold.
(i) *In* $A_2(\mathbb{D}^*)$ *topology,*

$$\lim_{n\to\infty} \mathcal{S}(g_n) = \mathcal{S}(g).$$

(ii) *In* $A_2^1(\mathbb{D}^*)$ *topology,*

$$\lim_{n\to\infty} \mathcal{A}(g_n) = \mathcal{A}(g).$$

PROOF. Since $\mathcal{T}_0(1)$ is a topological group, $\lim_{n\to\infty} \gamma_n^{-1} = \gamma^{-1} = \tilde{g}^{-1} \circ \tilde{f}$. We have $\tilde{f} = r \circ j \circ g \circ j$, so that
$$\mathcal{S}(\tilde{f}) = \overline{\mathcal{S}(g) \circ j} j_{\bar{z}}^2,$$
which proves part (i). Part (ii) follows from Lemma A.5. □

APPENDIX B

The Period Mapping $\hat{\mathscr{P}}$

Let $\mathscr{S}_\infty$ be the closed ideal of compact operators in the Banach algebra $\mathscr{B}(\ell^2)$ of bounded operators on ℓ^2. Here we prove that the period mapping $\hat{\mathscr{P}} : T(1) \to \mathscr{B}(\ell^2)$, defined in Remark 2.11, is a holomorphic mapping of complex Banach manifolds and that
$$\hat{\mathscr{P}}^{-1}(\mathscr{S}_\infty) = S = \text{Möb}(S^1)\backslash\text{Homeo}_s(S^1).$$

THEOREM B.1. *The inclusion* $\hat{\mathscr{P}} : T(1) \to \mathscr{B}(\ell^2)$ *is a holomorphic mapping of Banach manifolds.*

PROOF. As in the proof of Theorem 2.10, we will show that for every $[\nu] \in T(1)$ and $\mu \in \Omega^{-1,1}(\mathbb{D}^*)$, the map $\mathbb{C} \ni t \mapsto B_1(t) = B_1(f^{\nu+t\mu})$ is holomorphic in a neighborhood of $t = 0$ in $\mathbb{C}$. Choose $\delta > 0$ so that $\|\nu + t\mu\|_\infty < 1$ for all $|t| < \delta$. For every t_0 such that $|t_0| < \delta$, let δ_1 be such that $0 < \delta_1 < \delta - |t_0|$. Then for all $|t - t_0| < \delta_1$, we have as in Theorem 2.10,

$$\left(K_1^{\nu+t\mu} - K_1^{\nu+t_0\mu} - (t-t_0)\frac{d}{dt}\bigg|_{t=t_0} K_1^{\nu+t\mu} \right)(z,w)$$
$$= \frac{(t-t_0)^2}{2\pi i} \oint_{|\zeta-t_0|=\delta_1} \frac{K_1^{\nu+\zeta\mu}(z,w)}{(\zeta-t)(\zeta-t_0)^2} d\zeta.$$

This gives

$$\left\| \frac{B_1(f^{\nu+t\mu}) - B_1(f^{\nu+t_0\mu})}{t-t_0} - \frac{d}{dt}\bigg|_{t=t_0} B_1(f^{\nu+t\mu}) \right\|$$

$$= \sup_{\|u\|_2=1} \left(\iint_\mathbb{D} \left| \iint_\mathbb{D} \frac{t-t_0}{2\pi i} \oint_{|\zeta-t_0|=\delta_1} \frac{K_1^{\nu+\zeta\mu}(z,w)\overline{u(w)}}{(\zeta-t)(\zeta-t_0)^2} d\zeta d^2w \right|^2 d^2z \right)^{1/2}$$

$$\leq \frac{|t-t_0|}{2\pi} \sup_{\|u\|_2=1} \left(\iint_\mathbb{D} \left(\oint_{|\zeta-t_0|=\delta_1} \frac{|d\zeta|}{|\zeta-t|^2|\zeta-t_0|^4} \right) \right.$$
$$\left. \left(\oint_{|\zeta-t_0|=\delta_1} \left| \iint_\mathbb{D} K_1^{\nu+\zeta\mu}(z,w)\overline{u(w)}d^2w \right|^2 |d\zeta| \right) d^2z \right)^{1/2}$$

$$= \frac{|t-t_0|}{2\pi} \left(\oint_{|\zeta-t_0|=\delta_1} \frac{|d\zeta|}{|\zeta-t|^2|\zeta-t_0|^4} \right)^{1/2} \sup_{\|u\|_2=1} \left(\oint_{|\zeta-t_0|=\delta_1} \|K_1^{\nu+\zeta\mu}\bar{u}\|_2^2 |d\zeta| \right)^{1/2}.$$

Since $\|K_1\| < 1$, we obtain

$$\left\| \frac{B_1(f^{\nu+t\mu}) - B_1(f^{\nu+t_0\mu})}{t - t_0} - \frac{d}{dt}\bigg|_{t=t_0} B_1(f^{\nu+t\mu}) \right\|$$

$$\leq \frac{|t - t_0|}{2\pi} \left(\oint_{|\zeta-t_0|=\delta_1} \frac{|d\zeta|}{|\zeta - t|^2 |\zeta - t_0|^4} \oint_{|\zeta-t_0|=\delta_1} |d\zeta| \right)^{1/2}$$

$$= O(t - t_0) \quad \text{as} \quad t \to t_0.$$

□

To prove that $\hat{\mathscr{P}}(S) \subset \mathscr{S}_\infty$, we first give a characterization of the submanifold $S = \text{Möb}(S^1)\backslash\text{Homeo}_s(S^1)$ of $T(1)$. It has been shown by Gardiner and Sullivan [**GS92**] that $\beta(S) = A^0_\infty(\mathbb{D}) \cap \beta(T(1))$, where $\beta : T(1) \to A_\infty(\mathbb{D})$ is the Bers embedding and $A^0_\infty(\mathbb{D})$ is the subspace of the Banach space $A_\infty(\mathbb{D})$, defined by

$$A^0_\infty(\mathbb{D}) = \left\{ \phi \in A_\infty(\mathbb{D}) : \lim_{|z| \to 1^-} (1 - |z|^2)^2 \phi(z) = 0 \right\}.$$

Analogous to Theorem A.1 in Part I, we have the following result.

LEMMA B.2. *The closure of the homogeneous space* $\text{Möb}(S^1)\backslash\text{Diff}_+(S^1) \subset T(1)$ *in the Banach manifold topology is the Banach submanifold S of $T(1)$.*

PROOF. For $\phi \in A^0_\infty(\mathbb{D}) \cap \beta(T(1))$, let $\phi_n = \phi \circ r_n$, where r_n is the dilation $z \mapsto \frac{n}{n+1}z$, $n \in \mathbb{N}$. Since $\phi \in A^0_\infty(\mathbb{D})$, for every $\varepsilon > 0$ there exists $0 < r < 1$ such that

$$\sup_{r \leq |z| \leq 1} (1 - |z|^2)^2 |\phi(z)| < \frac{\varepsilon}{4}.$$

Thus there exists N' such that

$$\sup_{r' \leq |z| \leq 1} (1 - |z|^2)^2 |\phi_n(z)| < \frac{\varepsilon}{4}$$

for $n > N'$, where $r' = \frac{1+r}{2}$. The sequence $\{\phi_n\}$ converges uniformly to ϕ on compact subsets of $\mathbb{D}$, so that there exists N'' such that

$$\sup_{|z| \leq r'} (1 - |z|^2)^2 |\phi_n(z) - \phi(z)| < \frac{\varepsilon}{2} \quad \text{for} \quad n > N''.$$

Thus $\|\phi_n - \phi\|_\infty < \varepsilon$ for $n > N = \max\{N', N''\}$, so that

$$\lim_{n \to \infty} \phi_n = \phi$$

in the $A_\infty(\mathbb{D})$ topology. Since $\beta(T(1))$ is open in $A_\infty(\mathbb{D})$, $\phi_n \in \beta(T(1))$ for large enough n. The functions ϕ_n are smooth on S^1 (in fact analytic), so that corresponding $\gamma_n \in \text{Möb}(S^1)\backslash\text{Homeo}_{qs}(S^1)$ are also smooth on S^1. This proves that $\overline{\text{Möb}(S^1)\backslash\text{Diff}_+(S^1)} = S$. □

REMARK B.3. Together with Theorem A.1 in Part I, Lemma B.2 explains the distinguished role of the embedded manifold $\text{Möb}(S^1)\backslash\text{Diff}_+(S^1) \hookrightarrow T(1)$ in Teichmüller theory. Its closure in $T(1)$ under the Banach manifold topology is the Banach submanifold S, whereas its closure under the Hilbert manifold topology is the Hilbert submanifold $T_0(1)$.

THEOREM B.4. *The image of the Banach submanifold S under the KYNS period mapping $\hat{\mathscr{P}} : T(1) \to \mathscr{B}(\ell^2)$ is given by*

$$\hat{\mathscr{P}}(S) = \mathscr{S}_\infty \cap \hat{\mathscr{P}}(T(1)),$$

where $\mathscr{S}_\infty$ is the space of compact operators on ℓ^2.

PROOF. It is easy to show that $\hat{\mathscr{P}}(S) \subset \mathscr{S}_\infty$. Indeed, by Theorem 2.6, $\hat{\mathscr{P}}(\text{Möb}(S^1)\backslash \text{Diff}_+(S^1)) \subset \mathscr{S}_2 \subset \mathscr{S}_\infty$. Since the mapping $\hat{\mathscr{P}}$ is continuous (actually, holomorphic), using Lemma B.2 proves the claim.

To prove the converse inclusion $\hat{\mathscr{P}}^{-1}(\mathscr{S}_\infty \cap \hat{\mathscr{P}}(T(1))) \subset S$, we use methods developed by Brazilevic in [**Bra65**]. Let $\mathcal{U}$ be the space of univalent functions on $\mathbb{D}$. Following [**Bra65**], consider the following function $\mathsf{F} : \mathcal{U} \times \mathcal{U} \times \mathbb{D} \to \mathbb{R}$,

$$\mathsf{F}(f_1, f_2)(z) = \sqrt{\pi}(1 - |z|^2) \left(\iint_\mathbb{D} |K_1(f_1)(z,w) - K_1(f_2)(z,w)|^2 \, d^2 w \right)^{1/2}.$$

When $f_1 = f$ and $f_2 = \text{id}$ — the identity mapping, we denote

$$\mathsf{F}(f)(z) = \mathsf{F}(f, \text{id})(z) = \sqrt{\pi}(1 - |z|^2) \mathsf{K}_1(z,z)^{1/2}.$$

In [**Bra65**], Brazilevic has introduced a new metric on $\mathcal{U}$,

$$d(f_1, f_2) = \sup_{z \in \mathbb{D}} \mathsf{F}(f_1, f_2)(z),$$

and has shown that

$$\|\mathcal{S}(f_1) - \mathcal{S}(f_2)\|_\infty \leq 6 d(f_1, f_2).$$

For fixed $\zeta \in \mathbb{D}$, consider the kernel

$$K_1(f)(z, \zeta) = \frac{1}{\pi} \sum_{n=1}^{\infty} n \left(\sum_{m=1}^{\infty} m b_{-n,-m} \zeta^{m-1} \right) z^{n-1}$$

as a holomorphic function on $\mathbb{D}$. By Grunsky inequality,

$$\|K_1(f)(\,\cdot\,,\zeta)\|_2^2 = \mathsf{K}_1(f)(\zeta,\zeta) = \frac{1}{\pi} \sum_{n=1}^{\infty} n \left| \sum_{m=1}^{\infty} m b_{-n,-m} \zeta^{m-1} \right|^2$$

$$\leq \frac{1}{\pi} \sum_{n=1}^{\infty} n |\zeta|^{2n-2} = \frac{1}{\pi(1 - |\zeta|^2)^2} < \infty,$$

so that $K_1(f)(\,\cdot\,,\zeta) \in A_2^1(\mathbb{D})$. For fixed $\zeta \in \mathbb{D}$ and $f_1, f_2 \in \mathcal{U}$ we define

$$\psi(f_1, f_2; \zeta)(z) = K_1(f_1)(z, \zeta) - K_1(f_2)(z, \zeta).$$

Then $\psi(f_1, f_2, \zeta) \in A_2^1(\mathbb{D})$. For $\psi(f_1, f_2; \zeta) \neq 0$ we set

$$\mathsf{u}(f_1, f_2; \zeta) = \frac{\psi(f_1, f_2; \zeta)}{\|\psi(f_1, f_2; \zeta)\|_2},$$

and for $\psi(f_1, f_2, \zeta) = 0$ we set $\mathsf{u}(f_1, f_2; \zeta) = 0$. The following lemma generalizes Brazilevic's result [**Bra65**].

LEMMA B.5. *For $f_1, f_2 \in \mathcal{U}$ and $z \in \mathbb{D}$,*

$$(1 - |z|^2)^2 |\mathcal{S}(f_1)(z) - \mathcal{S}(f_2)(z)| \leq 6 \mathsf{F}(f_1, f_2)(z) \leq 6 \|K_1(f_1) - K_1(f_2)\|.$$

PROOF. We use the same approach as in [**Bra65**]. Since for $\lambda_1, \lambda_2 \in \mathrm{PSL}(2,\mathbb{C})$, $\mathcal{S}(\lambda_1 \circ f) = \mathcal{S}(f)$, $K_1(\lambda_1 \circ f) = K_1(f)$, and $\mathsf{F}(\lambda_1 \circ f_1, \lambda_2 \circ f_2)(z) = \mathsf{F}(f_1, f_2)(z)$, it is sufficient to consider only $f \in \mathcal{U}$ normalized by $f(0) = 0$ and $f'(0) = 1$. We have for fixed $z \in \mathbb{D}$,

$$(1 - |z|^2)^2 \mathcal{S}(f)(z) = \mathcal{S}(\lambda(f)_z \circ f \circ \sigma_z)(0)$$

and

$$\mathsf{F}(f_1, f_2)(z) = \mathsf{F}(\lambda(f_1)_z \circ f_1 \circ \sigma_z, \lambda(f_2)_z \circ f_2 \circ \sigma_z)(0),$$

where $\sigma_z \in \mathrm{M\"ob}(S^1)$ and $\lambda(f)_z \in \mathrm{PSL}(2,\mathbb{C})$ are given by[1]

$$\sigma_z(w) = \frac{w + z}{1 + w\bar{z}} \quad \text{and} \quad \lambda(f)_z(w) = \frac{w - f(z)}{f'(z)(1 - |z|^2)}.$$

Since for a normalized $f \in \mathcal{U}$ the univalent function $\lambda(f)_z \circ f \circ \sigma_z$ is also normalized, for the first inequality we need only to show that for any normalized $f \in \mathcal{U}$,

$$|\mathcal{S}(f_1)(0) - \mathcal{S}(f_2)(0)| \leq 6 \mathsf{F}(f_1, f_2)(0).$$

Since

$$\mathcal{S}(f)(z) = 6 \lim_{w \to z} \left(\frac{f'(z)f'(w)}{(f(z) - f(w))^2} - \frac{1}{(z-w)^2} \right) = -6 \sum_{n,m=1}^{\infty} nm b_{-n,-m} z^{n+m-2},$$

we have

$$|\mathcal{S}(f_1)(0) - \mathcal{S}(f_2)(0)| = 6|b_{-1,-1}(f_1) - b_{-1,-1}(f_2)|.$$

On the other hand, it is straightforward to compute that

$$\mathsf{F}(f_1, f_2)^2(0) = \pi \iint_{\mathbb{D}} |K_1(f_1)(0,w) - K_1(f_2)(0,w)|^2 d^2w$$

$$= \sum_{m=1}^{\infty} m|b_{-1,-m}(f_1) - b_{-1,-m}(f_2)|^2,$$

and the first inequality follows.

Next we observe that

(B.1) $$\mathsf{F}(f_1, f_2)(z) = \sqrt{\pi} \|(K_1(f_1) - K_1(f_2))\overline{\mathsf{u}(f_1, f_2; z)}\|_{A^1_\infty(\mathbb{D})}.$$

Indeed, by Cauchy-Schwarz inequality,

$$\left((K_1(f_1) - K_1(f_2)) \overline{\psi(f_1, f_2; z)} \right)(w)$$

$$= \iint_{\mathbb{D}} (K_1(f_1)(w, \zeta) - K_1(f_2)(w, \zeta)) \overline{(K_1(f_1)(\zeta, z) - K_1(f_2)(\zeta, z))} d^2\zeta$$

$$\leq \|\psi(f_1, f_2; z)\|_2 \|\psi(f_1, f_2; w)\|_2,$$

---

[1]Here subscript z does not denote a derivative.

with the equality for $w = z$. Hence

$$\left\| (K_1(f_1) - K_1(f_2))\overline{\psi(f_1, f_2; z)} \right\|_{A_\infty^1(\mathbb{D})}$$
$$= (1 - |z|^2) \iint_{\mathbb{D}} |K_1(f_1)(\zeta, z) - K_1(f_2)(\zeta, z)|^2 \, d^2\zeta$$
$$= (1 - |z|^2)\|\psi(f_1, f_2; z)\|_2^2 = \mathsf{F}(f_1, f_2)(z) \frac{\|\psi(f_1, f_2; z)\|_2}{\sqrt{\pi}}.$$

Finally, using (B.1) and the estimate in Lemma 1.3, we get

(B.2) $\quad \mathsf{F}(f_1, f_2)(z) \leq \|(K_1(f_1) - K_1(f_2))\overline{\mathsf{u}(f_1, f_2; z)}\|_2 \leq \|K_1(f_1) - K_1(f_2)\|.$

$\square$

REMARK B.6. It immediately follows from Lemma B.5 that

$$\|\mathcal{S}(f_1) - \mathcal{S}(f_2)\|_\infty \leq 6d(f_1, f_2) \leq 6\|K_1(f_1) - K_1(f_2)\|,$$

which is a stronger version of Brazilevic's result [**Bra65**]. In case $f_1 = f$ and $f_2 = \mathrm{id}$ we have

$$\|\mathcal{S}(f)\|_\infty \leq 6d(f) \leq 6\|K_1(f)\|,$$

where $d(f) = d(f, \mathrm{id})$. Since $\|K_1(f)\| \leq 1$, where equality holds if and only if $\mathbb{C} \setminus f(\mathbb{D})$ has Lebesgue measure zero, this recovers another result in [**Bra65**] that $d(f) \leq 1$ for $f \in \mathcal{U}$, and $d(f) = 1$ implies that $\mathbb{C} \setminus f(\mathbb{D})$ has Lebesgue measure zero.

Given a normalized univalent function $f : \mathbb{D} \to \mathbb{C}$, let $f_n : \mathbb{D} \to \mathbb{C}$ be the normalized univalent function defined by $f_n = r_n^{-1} \circ f \circ r_n$, where r_n is the dilation $z \mapsto \frac{n}{n+1} z$. Since f_n is analytic on S^1, we have

$$\lim_{|z| \to 1^-} (1 - |z|^2)^2 \mathcal{S}(f_n)(z) = 0,$$
$$\lim_{|\zeta| \to 1^-} (1 - |\zeta|^2) K_1(f_n)(z, \zeta) = 0,$$

and also

$$\lim_{|\zeta| \to 1^-} \|(1 - |\zeta|^2) K_1(f_n)(\,\cdot\,, \zeta)\|_2^2 = \lim_{|\zeta| \to 1^-} (1 - |\zeta|^2)^2 \mathsf{K}_1(f_n)(\zeta, \zeta) = 0.$$

LEMMA B.7. *Let $f : \mathbb{D} \to \mathbb{C}$ be a normalized univalent function and let $\{f_n\}_{n=1}^\infty$ be the sequence of normalized univalent functions defined above. Then*

$$\lim_{n \to \infty} K_1(f_n) = K_1(f)$$

in the strong operator topology.

PROOF. For $\psi \in A_2^1(\mathbb{D})$ set $\psi_n = r_n \circ \psi \circ r_n$. It is elementary to show that

$$\lim_{n \to \infty} \|\psi - \psi_n\|_2 = 0.$$

For $(K_1(f)\bar\psi)_n = r_n \circ (K_1(f)\bar\psi) \circ r_n$ we have,

$$K_1(f_n)\bar\psi_n = (K_1(f)\bar\psi)_n - r_n \circ (K_1(f)\overline{\psi(1 - \chi_n)}) \circ r_n,$$

where χ_n is the characteristic function of the disk $\mathbb{D}_n = r_n(\mathbb{D})$. Using this identity and the inequalities $\|K_1(f)\| \leq 1$, $\|\psi_n\|_2 \leq \|\psi\|_2$, we obtain

$$\|(K_1(f) - K_1(f_n))\bar{\psi}\|_2 \leq \|K_1(f)\bar{\psi} - K_1(f_n)\bar{\psi}_n\|_2 + \|K_1(f_n)\overline{(\psi_n - \psi)}\|_2$$
$$\leq \|K_1(f)\bar{\psi} - (K_1(f)\bar{\psi})_n\|_2 + \|K_1(f)\overline{(\psi(1 - \chi_n))}\|_2 + \|\psi - \psi_n\|_2$$
$$\leq \|K_1(f)\bar{\psi} - (K_1(f)\bar{\psi})_n\|_2 + \|\psi(1 - \chi_n)\|_2 + \|\psi - \psi_n\|_2.$$

Since $\psi \in A_2^1(\mathbb{D})$,
$$\lim_{n \to \infty} \|\psi(1 - \chi_n)\|_2 = 0,$$
and we get the assertion of the lemma. $\square$

LEMMA B.8. *Let $\gamma = g^{-1} \circ f \in T(1)$ be such that $K_1(f)$ is a compact operator. Then for every sequence $\{\zeta_m\}_{m=1}^\infty$ of points in $\mathbb{D}$, the corresponding sequence of functions $\{u_m\}_{m=1}^\infty$ in $A_2^1(\mathbb{D})$, where*
$$u_m(z) = (1 - |\zeta_m|^2)K_1(f)(z, \zeta_m), \ z \in \mathbb{D},$$
contains a convergent subsequence in $A_2^1(\mathbb{D})$.

PROOF. Consider the following sequence of functions,
$$v_m(z) = z^{-2}(1 - |\zeta_m|^2)K_3(f)(z^{-1}, \zeta_m) \in A_2^1(\mathbb{D}).$$
Using the formula
$$\mathsf{K}_3(\zeta, \zeta) + \mathsf{K}_4(\zeta, \zeta) = \frac{1}{\pi(1 - |\zeta|^2)^2},$$
which follows from the operator identity $\mathsf{K}_3 + \mathsf{K}_4 = I$, and the inequality $\mathsf{K}_4(\zeta, \zeta) \geq 0$, we get
$$\|v_m\|_2^2 = (1 - |\zeta_m|^2)^2 \mathsf{K}_3(\zeta_m, \zeta_m) \leq \frac{1}{\pi}.$$
Now consider the operator $\tilde{K}_3(f) : \overline{A_2^1(\mathbb{D})} \to A_2^1(\mathbb{D})$, defined by the kernel
$$\tilde{K}_3(f)(z, w) = z^{-2} K_3(f)\left(z^{-1}, w\right).$$
In the standard basis for $A_2^1(\mathbb{D})$ it is given by the matrix $B_3(f)$ and, therefore, is a topological isomorphism. Setting $K(f) = K_1(f)\tilde{K}_3(f)^{-1}$, we get
$$u_m = K(f)v_m.$$
Since the operator $K(f)$ is compact and the sequence $\{v_m\}_{m=1}^\infty$ is bounded, the statement follows. $\square$

Now we can finish the proof of the Theorem. Suppose that for $[\mu] \in T(1)$ the corresponding operator $K_1(f)$ is compact but $[\mu] \notin S$. According to Remark 1.7, this implies that there exist $\varepsilon > 0$ and a sequence $\zeta_m \in \mathbb{D}$ satisfying
$$|\zeta_m| > 1 - \frac{1}{m} \quad \text{and} \quad (1 - |\zeta_m|^2)^2|\mathcal{S}(f)(\zeta_m)| \geq \varepsilon.$$
By Lemma B.8, there exists a subsequence ζ_{m_k} such that the sequence of functions
$$u_{m_k}(z) = (1 - |\zeta_{m_k}|^2)K_1(f)(z, \zeta_{m_k})$$
converges to $u \in A_2^1(\mathbb{D})$ in $A_2^1(\mathbb{D})$. Since
$$\lim_{|\zeta| \to 1^-} (1 - |\zeta|^2)K_1(f_n)(z, \zeta) = 0,$$

for any $n \in \mathbb{N}$, the sequence of functions
$$(1 - |\zeta_{m_k}|^2)\psi(f, f_n; \zeta_{m_k}) = (1 - |\zeta_{m_k}|^2)\left(K_1(f)(\,\cdot\,, \zeta_{m_k}) - K_1(f_n)(\,\cdot\,, \zeta_{m_k})\right)$$
also converges to u as $k \to \infty$. From Lemma B.5 and (B.2) we get the following inequality
$$(1 - |\zeta_{m_k}|^2)^2 \left|\mathcal{S}(f)(\zeta_{m_k}) - \mathcal{S}(f_n)(\zeta_{m_k})\right| \leq 6 \left\|(K_1(f) - K_1(f_n))\overline{u(f, f_n; \zeta_{m_k})}\right\|_2,$$
which for $\psi(f, f_n, \zeta_{m_k}) = 0$ is an equality. Now passing to the limit $k \to \infty$ for fixed $n \in \mathbb{N}$, we obtain
$$\varepsilon \leq 6 \left\|(K_1(f) - K_1(f_n))\bar{\mathrm{u}}\right\|_2,$$
where
$$\mathrm{u} = \frac{u}{\|u\|_2} \neq 0.$$
However, according to Lemma B.7,
$$\lim_{n \to \infty} \left\|(K_1(f) - K_1(f_n))\bar{\mathrm{u}}\right\|_2 = 0.$$
This contradiction proves that $[\mu] \in S$. □

REMARK B.9. For $[\mu] \in S$ the proof of Lemma B.2 shows that
$$\lim_{n \to \infty} \mathcal{S}(f_n) = \mathcal{S}(f)$$
in $A_\infty(\mathbb{D})$ topology. Since the period mapping $\hat{\mathscr{P}}$ is continuous,
$$\lim_{n \to \infty} K_1(f_n) = K_1(f)$$
in the norm topology on $\mathscr{B}(\overline{A_2^1(\mathbb{D})}, A_2^1(\mathbb{D}))$.

The following commutative diagram displays the properties of the tower of embedded manifolds $T_0(1) \hookrightarrow S \hookrightarrow T(1)$ under the KYNS period mapping $\hat{\mathscr{P}}$, the pre-Bers embedding $\hat{\beta}$ and the Bers embedding $\beta = \Psi \circ \hat{\beta}$,

$$\begin{array}{ccccc}
\mathscr{S}_2(\ell^2) & \longrightarrow & \mathscr{S}_\infty(\ell^2) & \longrightarrow & \mathscr{B}(\ell^2) \\
\uparrow \mathscr{P} & & \uparrow \hat{\mathscr{P}} & & \uparrow \hat{\mathscr{P}} \\
T_0(1) & \longrightarrow & S & \longrightarrow & T(1) \\
\downarrow \hat{\beta} & & \downarrow \hat{\beta} & & \downarrow \hat{\beta} \\
A_2^1(\mathbb{D}) & \longrightarrow & A_\infty^{1,0}(\mathbb{D}) & \longrightarrow & A_\infty^1(\mathbb{D}) \\
\downarrow \Psi & & \downarrow \Psi & & \downarrow \Psi \\
A_2(\mathbb{D}) & \longrightarrow & A_\infty^0(\mathbb{D}) & \longrightarrow & A_\infty(\mathbb{D})
\end{array}$$

Here $A_\infty^{1,0}(\mathbb{D})$ is the closed subspace of $A_\infty^1(\mathbb{D})$, defined by
$$A_\infty^{1,0}(\mathbb{D}) = \left\{\psi \in A_\infty^1(\mathbb{D}) : \lim_{|z| \to 1^-}(1 - |z|^2)\psi(z) = 0\right\}.$$
All horizontal maps are embeddings, and all vertical maps are holomorphic mappings of Banach and Hilbert manifolds respectively. All these properties have been

proved already, except for the simple fact $\Psi(A^{1,0}_\infty(\mathbb{D})) \subset A^0_\infty(\mathbb{D})$, which easily follows from Cauchy integral formula.

Bibliography

[AB60] Lars Ahlfors and Lipman Bers, *Riemann's mapping theorem for variable metrics*, Ann. of Math. (2) **72** (1960), 385–404.

[Ahl61] Lars V. Ahlfors, *Some remarks on Teichmüller's space of Riemann surfaces*, Ann. of Math. (2) **74** (1961), 171–191.

[Ahl62] _____, *Curvature properties of Teichmüller's space*, J. Analyse Math. **9** (1961/1962), 161–176.

[Ahl87] _____, *Lectures on quasiconformal mappings*, Wadsworth & Brooks/Cole Advanced Books & Software, Monterey, CA, 1987, With the assistance of Clifford J. Earle, Jr., Reprint of the 1966 original.

[Ber65] L Bers, *Automorphic forms and general Teichmüller spaces*, Proc. Conf. Complex Analysis (Minneapolis, 1964), Springer, Berlin, 1965, pp. 109–113.

[Ber66] Lipman Bers, *A non-standard integral equation with applications to quasiconformal mappings*, Acta Math. **116** (1966), 113–134.

[Ber72] _____, *Uniformization, moduli and Kleinian groups*, Bull. London. Math. Soc. **4** (1972), 257–300.

[Ber73] _____, *Fiber spaces over Teichmüller spaces*, Acta. Math. **130** (1973), 89–126.

[BFK92] D. Burghelea, L. Friedlander, and T. Kappeler, *Meyer-Vietoris type formula for determinants of elliptic differential operators*, J. Funct. Anal. **107** (1992), no. 1, 34–65.

[Bou67] N. Bourbaki, *Éléments de mathématique. Fasc. XXXIII. Variétés différentielles et analytiques. Fascicule de résultats (Paragraphes 1 à 7)*, Actualités Scientifiques et Industrielles, No. 1333, Hermann, Paris, 1967.

[BP78] Jochen Becker and Christian Pommerenke, *Über die quasikonforme Fortsetzung schlichter Funktionen*, Math. Z. **161** (1978), no. 1, 69–80.

[BR87a] M. J. Bowick and S. G. Rajeev, *The holomorphic geometry of closed bosonic string theory and* Diff S^1/S^1, Nuclear Phys. B **293** (1987), no. 2, 348–384.

[BR87b] _____, *String theory as the Kähler geometry of loop space*, Phys. Rev. Lett. **58** (1987), no. 6, 535–538.

[Bra65] J. E. Brazilevič, *On dispersion of coefficients of univalent functions*, Mat. Sb. (N.S.) **68 (110)** (1965), 549–560.

[DE86] Adrien Douady and Clifford J. Earle, *Conformally natural extension of homeomorphisms of the circle*, Acta Math. **157** (1986), no. 1-2, 23–48.

[Dur83] P. L. Duren, *Univalent functions*, Grundlehren der Mathematischen Wissenschaften [Fundamental Principles of Mathematical Sciences], vol. 259, Springer-Verlag, New York, 1983.

[EN88] Clifford J. Earle and Subhashis Nag, *Conformally natural reflections in Jordan curves with applications to Teichmüller spaces*, Holomorphic functions and moduli, Vol. II (Berkeley, CA, 1986), Math. Sci. Res. Inst. Publ., vol. 11, Springer, New York, 1988, pp. 179–194.

[GK69] I. C. Gohberg and M. G. Kreĭn, *Introduction to the theory of linear nonselfadjoint operators*, Translated from the Russian by A. Feinstein. Translations of Mathematical Monographs, Vol. 18, American Mathematical Society, Providence, R.I., 1969.

[GS92] Frederick P. Gardiner and Dennis P. Sullivan, *Symmetric structures on a closed curve*, Amer. J. Math. **114** (1992), no. 4, 683–736.

[Ham02] D. H. Hamilton, *Conformal welding*, Handbook of complex analysis: geometric function theory, Vol. 1, North-Holland, Amsterdam, 2002, pp. 137–146.

[Hej76] Dennis A. Hejhal, *The Selberg trace formula for* PSL(2, R). *Vol. I*, Springer-Verlag, Berlin, 1976, Lecture Notes in Mathematics, Vol. 548.

[Hum72] J. A. Hummel, *Inequalities of Grunsky type for Aharonov pairs*, J. Analyse Math. **25** (1972), 217–257.

[HZ99] Andrew Hassell and Steve Zelditch, *Determinants of Laplacians in exterior domains*, Internat. Math. Res. Notices (1999), no. 18, 971–1004.

[Kir87] A. A. Kirillov, *Kähler structure on the K-orbits of a group of diffeomorphisms of the circle*, Funktsional. Anal. i Prilozhen. **21** (1987), no. 2, 42–45.

[KY87] A. A. Kirillov and D. V. Yur′ev, *Kähler geometry of the infinite-dimensional homogeneous space $M = \text{diff}_+(S^1)/\text{rot}(S^1)$*, Funktsional. Anal. i Prilozhen. **21** (1987), no. 4, 35–46.

[KY88] A. A. Kirillov and D. V. Yuriev, *Representations of the Virasoro algebra by the orbit method*, J. Geom. Phys. **5** (1988), no. 3, 351–363.

[Lan85] Serge Lang, $SL_2(\mathbf{R})$, Graduate Texts in Mathematics, vol. 105, Springer-Verlag, New York, 1985, Reprint of the 1975 edition.

[Lan95] _____, *Differential and Riemannian manifolds*, third ed., Springer-Verlag, New York, 1995.

[Leh87] Olli Lehto, *Univalent functions and Teichmüller spaces*, Springer-Verlag, New York, 1987.

[Nag88] Subhashis Nag, *The complex analytic theory of Teichmüller spaces*, John Wiley & Sons Inc., New York, 1988, A Wiley-Interscience Publication.

[Nag92] _____, *A period mapping in universal Teichmüller space*, Bull. Amer. Math. Soc. (N.S.) **26** (1992), no. 2, 280–287.

[NS95] Subhashis Nag and Dennis Sullivan, *Teichmüller theory and the universal period mapping via quantum calculus and the $H^{1/2}$ space on the circle*, Osaka J. Math. **32** (1995), no. 1, 1–34.

[NV90] Subhashis Nag and Alberto Verjovsky, $\text{diff}(S^1)$ *and the Teichmüller spaces*, Comm. Math. Phys. **130** (1990), no. 1, 123–138.

[Pat75] S. J. Patterson, *A lattice-point problem in hyperbolic space*, Mathematika **22** (1975), no. 1, 81–88.

[Pom92] Ch. Pommerenke, *Boundary behaviour of conformal maps*, Springer-Verlag, Berlin, 1992.

[PS86] Andrew Pressley and Graeme Segal, *Loop groups*, Oxford Mathematical Monographs, The Clarendon Press Oxford University Press, New York, 1986.

[Roy75] H. L. Royden, *Intrinsic metrics on Teichmüller space*, Proceedings of the International Congress of Mathematicians (Vancouver, B. C., 1974), Vol. 2, Canad. Math. Congress, Montreal, Que., 1975, pp. 217–221.

[Sch57] M. Schiffer, *The Fredholm eigen values of plane domains*, Pacific J. Math. **7** (1957), 1187–1225.

[Sch59] _____, *Fredholm eigen values of multiply-connected domains*, Pacific J. Math. **9** (1959), 211–269.

[Sch81] Menahem Schiffer, *Fredholm eigenvalues and Grunsky matrices*, Ann. Polon. Math. **39** (1981), 149–164.

[Seg81] Graeme Segal, *Unitary representations of some infinite-dimensional groups*, Comm. Math. Phys. **80** (1981), no. 3, 301–342.

[SH62] M. Schiffer and N. S. Hawley, *Connections and conformal mapping*, Acta Math. **107** (1962), 175–274.

[Sie64] Carl Ludwig Siegel, *Symplectic geometry*, Academic Press, New York, 1964.

[STZ99] Maria E. Schonbek, Andrey N. Todorov, and Jorge P. Zubelli, *Geodesic flows on diffeomorphisms of the circle, Grassmannians, and the geometry of the periodic KdV equation*, Adv. Theor. Math. Phys. **3** (1999), no. 4, 1027–1092.

[SW85] Graeme Segal and George Wilson, *Loop groups and equations of KdV type*, Inst. Hautes Études Sci. Publ. Math. (1985), no. 61, 5–65.

[Teo03] Lee-Peng Teo, *Analytic functions and integrable hierarchies—characterization of tau functions*, Lett. Math. Phys. **64** (2003), no. 1, 75–92.

[Teo04] _____, *The Velling-Kirillov metric on the universal Teichmüller curve*, J. Analyse Math. **93** (2004), 271–308.

[TT03] Leon A. Takhtajan and Lee-Peng Teo, *Liouville action and Weil-Petersson metric on deformation spaces, global Kleinian reciprocity and holography*, Comm. Math. Phys. **239** (2003), no. 1-2, 183–240.

[TZ91] L. A. Takhtajan and P. G. Zograf, *A local index theorem for families of $\overline{\partial}$-operators on punctured Riemann surfaces and a new Kähler metric on their moduli spaces*, Comm. Math. Phys. **137** (1991), no. 2, 399–426.

[Vel] John A. Velling, *A projectively natural metric on Teichmüller's spaces*, unpublished manuscript.

[Wol86] Scott A. Wolpert, *Chern forms and the Riemann tensor for the moduli space of curves*, Invent. Math. **85** (1986), no. 1, 119–145.

[YB53] K. Yano and S. Bochner, *Curvature and Betti numbers*, Annals of Mathematics Studies, No. 32, Princeton University Press, Princeton, N. J., 1953.

[ZT87] P. G. Zograf and L. A. Takhtadzhyan, *A local index theorem for families of $\overline{\partial}$-operators on Riemann surfaces*, Uspekhi Mat. Nauk **42** (1987), no. 6(258), 133–150.

Editorial Information

To be published in the *Memoirs*, a paper must be correct, new, nontrivial, and significant. Further, it must be well written and of interest to a substantial number of mathematicians. Piecemeal results, such as an inconclusive step toward an unproved major theorem or a minor variation on a known result, are in general not acceptable for publication. Papers appearing in *Memoirs* are generally at least 80 and not more than 200 published pages in length. Papers less than 80 or more than 200 published pages require the approval of the Managing Editor of the Transactions/Memoirs Editorial Board.

As of May 31, 2006, the backlog for this journal was approximately 11 volumes. This estimate is the result of dividing the number of manuscripts for this journal in the Providence office that have not yet gone to the printer on the above date by the average number of monographs per volume over the previous twelve months, reduced by the number of volumes published in four months (the time necessary for preparing a volume for the printer). (There are 6 volumes per year, each containing at least 4 numbers.)

A Consent to Publish and Copyright Agreement is required before a paper will be published in the *Memoirs*. After a paper is accepted for publication, the Providence office will send a Consent to Publish and Copyright Agreement to all authors of the paper. By submitting a paper to the *Memoirs*, authors certify that the results have not been submitted to nor are they under consideration for publication by another journal, conference proceedings, or similar publication.

Information for Authors

Memoirs are printed from camera copy fully prepared by the author. This means that the finished book will look exactly like the copy submitted.

The paper must contain a *descriptive title* and an *abstract* that summarizes the article in language suitable for workers in the general field (algebra, analysis, etc.). The *descriptive title* should be short, but informative; useless or vague phrases such as "some remarks about" or "concerning" should be avoided. The *abstract* should be at least one complete sentence, and at most 300 words. Included with the footnotes to the paper should be the 2000 *Mathematics Subject Classification* representing the primary and secondary subjects of the article. The classifications are accessible from www.ams.org/msc/. The list of classifications is also available in print starting with the 1999 annual index of *Mathematical Reviews*. The Mathematics Subject Classification footnote may be followed by a list of *key words and phrases* describing the subject matter of the article and taken from it. Journal abbreviations used in bibliographies are listed in the latest *Mathematical Reviews* annual index. The series abbreviations are also accessible from www.ams.org/publications/. To help in preparing and verifying references, the AMS offers MR Lookup, a Reference Tool for Linking, at www.ams.org/mrlookup/. When the manuscript is submitted, authors should supply the editor with electronic addresses if available. These will be printed after the postal address at the end of the article.

Electronically prepared manuscripts. The AMS encourages electronically prepared manuscripts, with a strong preference for $\mathcal{AMS}$-LaTeX. To this end, the Society has prepared $\mathcal{AMS}$-LaTeX author packages for each AMS publication. Author packages include instructions for preparing electronic manuscripts, the *AMS Author Handbook*, samples, and a style file that generates the particular design specifications of that publication series. Though $\mathcal{AMS}$-LaTeX is the highly preferred format of TeX, author packages are also available in $\mathcal{AMS}$-TeX.

Authors may retrieve an author package from e-MATH starting from www.ams.org/tex/ or via FTP to ftp.ams.org (login as anonymous, enter username as password, and type cd pub/author-info). The *AMS Author Handbook* and the *Instruction Manual* are available in PDF format following the author packages link from www.ams.org/tex/. The author package can also be obtained free of charge by sending email to `tech-support@ams.org` (Internet) or from the Publication Division, American Mathematical Society, 201 Charles St., Providence, RI 02904-2294, USA. When requesting an author package, please specify $\mathcal{AMS}$-LaTeX or $\mathcal{AMS}$-TeX and the publication in which your paper will appear. Please be sure to include your complete mailing address.

Sending electronic files. After acceptance, the source file(s) should be sent to the Providence office (this includes any TeX source file, any graphics files, and the DVI or PostScript file).

Before sending the source file, be sure you have proofread your paper carefully. The files you send must be the EXACT files used to generate the proof copy that was accepted for publication. For all publications, authors are required to send a printed copy of their paper, which exactly matches the copy approved for publication, along with any graphics that will appear in the paper.

TeX files may be submitted by email, FTP, or on diskette. The DVI file(s) and PostScript files should be submitted only by FTP or on diskette unless they are encoded properly to submit through email. (DVI files are binary and PostScript files tend to be very large.)

Electronically prepared manuscripts can be sent via email to `pub-submit@ams.org` (Internet). The subject line of the message should include the publication code to identify it as a Memoir. TeX source files, DVI files, and PostScript files can be transferred over the Internet by FTP to the Internet node `e-math.ams.org` (130.44.1.100).

Electronic graphics. Comprehensive instructions on preparing graphics are available at `www.ams.org/jourhtml/graphics.html`. A few of the major requirements are given here.

Submit files for graphics as EPS (Encapsulated PostScript) files. This includes graphics originated via a graphics application as well as scanned photographs or other computer-generated images. If this is not possible, TIFF files are acceptable as long as they can be opened in Adobe Photoshop or Illustrator. No matter what method was used to produce the graphic, it is necessary to provide a paper copy to the AMS.

Authors using graphics packages for the creation of electronic art should also avoid the use of any lines thinner than 0.5 points in width. Many graphics packages allow the user to specify a "hairline" for a very thin line. Hairlines often look acceptable when proofed on a typical laser printer. However, when produced on a high-resolution laser imagesetter, hairlines become nearly invisible and will be lost entirely in the final printing process.

Screens should be set to values between 15% and 85%. Screens which fall outside of this range are too light or too dark to print correctly. Variations of screens within a graphic should be no less than 10%.

Inquiries. Any inquiries concerning a paper that has been accepted for publication should be sent directly to the Electronic Prepress Department, American Mathematical Society, 201 Charles St., Providence, RI 02904, USA.

Editors

This journal is designed particularly for long research papers, normally at least 80 pages in length, and groups of cognate papers in pure and applied mathematics. Papers intended for publication in the *Memoirs* should be addressed to one of the following editors. In principle the Memoirs welcomes electronic submissions, and some of the editors, those whose names appear below with an asterisk (*), have indicated that they prefer them. However, editors reserve the right to request hard copies after papers have been submitted electronically. Authors are advised to make preliminary email inquiries to editors about whether they are likely to be able to handle submissions in a particular electronic form.

*Algebra to ALEXANDER KLESHCHEV, Department of Mathematics, University of Oregon, Eugene, OR 97403-1222; email: ams@noether.uoregon.edu

Algebra and its application to MINA TEICHER, Emmy Noether Research Institute for Mathematics, Bar-Ilan University, Ramat-Gan 52900, Israel; email: teicher@macs.biu.ac.il

Algebraic geometry to DAN ABRAMOVICH, Department of Mathematics, Brown University, Box 1917, Providence, RI 02912; email: amsedit@math.brown.edu

*Algebraic number theory to V. KUMAR MURTY, Department of Mathematics, University of Toronto, 100 St. George Street, Toronto, ON M5S 1A1, Canada; email: murty@math.toronto.edu

*Algebraic topology to ALEJANDRO ADEM, Department of Mathematics, University of British Columbia, Room 121, 1984 Mathematics Road, Vancouver, British Columbia, Canada V6T 1Z2; email: adem@math.ubc.ca

*Combinatorics to JOHN R. STEMBRIDGE, Department of Mathematics, University of Michigan, Ann Arbor, Michigan 48109-1109; email: FRS@umich.edu

Complex analysis and harmonic analysis to ALEXANDER NAGEL, Department of Mathematics, University of Wisconsin, 480 Lincoln Drive, Madison, WI 53706-1313; email: nagel@math.wisc.edu

*Differential geometry and global analysis to LISA C. JEFFREY, Department of Mathematics, University of Toronto, 100 St. George St., Toronto, ON Canada M5S 3G3; email: jeffrey@math.toronto.edu

Dynamical systems and ergodic theory to AMIE WILKINSON, Department of Mathematics, Northwestern University, 2033 Sheridan Road, Evanston, IL 60208-2730; email: transactions@math.northwestern.edu

*Functional analysis and operator algebras to MARIUS DADARLAT, Department of Mathematics, Purdue University, 150 N. University St., West Lafayette, IN 47907-2067; email: mdd@math.purdue.edu

*Geometric analysis to TOBIAS COLDING, Courant Institute, New York University, 251 Mercer St., New York, NY 10012; email: traneditor@cims.nyu.edu

*Geometric analysis to MLADEN BESTVINA, Department of Mathematics, University of Utah, 155 South 1400 East, JWB 233, Salt Lake City, Utah 84112-0090; email: bestvina@math.utah.edu

Harmonic analysis, representation theory, and Lie theory to ROBERT J. STANTON, Department of Mathematics, The Ohio State University, 231 West 18th Avenue, Columbus, OH 43210-1174; email: stanton@math.ohio-state.edu

*Logic to STEFFEN LEMPP, Department of Mathematics, University of Wisconsin, 480 Lincoln Drive, Madison, Wisconsin 53706-1388; email: lempp@math.wisc.edu

*Ordinary differential equations, and applied mathematics to PETER W. BATES, Department of Mathematics, Michigan State University, East Lansing, MI 48824-1027; email: bates@math.msu.edu

*Partial differential equations to GUSTAVO PONCE, Department of Mathematics, South Hall, Room 6607, University of California, Santa Barbara, CA 93106; email: ponce@math.ucsb.edu

*Probability and statistics to KRZYSZTOF BURDZY, Department of Mathematics, University of Washington, Box 354350, Seattle, Washington 98195-4350; email: burdzy@math.washington.edu

*Real analysis and partial differential equations to DANIEL TATARU, Department of Mathematics, University of California, Berkeley, Berkeley, CA 94720; email: tataru@math.berkeley.edu

All other communications to the editors should be addressed to the Managing Editor, ROBERT GURALNICK, Department of Mathematics, University of Southern California, Los Angeles, CA 90089-1113; email: guralnic@math.usc.edu.

Titles in This Series

864 **John M. Lee,** Fredholm operators and Einstein metrics on conformally compact manifolds, 2006

863 **M. Lübke and A. Teleman,** The universal Kobayashi-Hitchin correspondence on Hermitian manifolds, 2006

862 **Alberto Canonaco,** The Beilinson complex and canonical rings of irregular surfaces, 2006

861 **Leon A. Takhtajan and Lee-Peng Teo,** Weil-Petersson metric on the universal Teichmüller space, 2006

860 **Thomas M. Fiore,** Pseudo limits, biadjoints and pseudo algebras: Categorical foundations of conformal field theory, 2006

859 **N. Arcozzi, R. Rochberg, and E. Sawyer,** Carleson measures and interpolating sequences for Besov spaces on complex balls, 2006

858 **Enrico Valdinoci, Berardino Sciunzi, and Vasile Ovidiu Savin,** Flat level set regularity of p-Laplace phase transitions, 2006

857 **Donatella Danielli, Nocola Garofalo, and Duy-Minh Nhieu,** Non-doubling Ahlfors measures, perimeter measures, and the characterization of the trace spaces of Sobolev functions in Carnot-Carathéodory spaces, 2006

856 **Vladimir Bolotnikov and Harry Dym,** On boundary interpolation for matrix valued Schur functions, 2006

855 **Yevgenia Kashina, Yorck Sommerhäuser, and Yongchang Zhu,** On higher Frobenius-Schur indicators, 2006

854 **Noam Greenberg,** The role of true finiteness in the admissible recursively enumerable degrees, 2006

853 **Joachim Krieger,** Stability of spherically symmetric wave maps, 2006

852 **Viorel Barbu, Irena Lasiecka, and Roberto Triggiani,** Tangential boundary stabilization of Navier-Stokes equations, 2006

851 **Jie Wu,** On maps from loop suspensions to loop spaces and the shuffle relations on the Cohen groups, 2006

850 **Siegfried Echterhoff, S. Kaliszewski, John Quigg, and Iain Raeburn,** A categorical approach to imprimitivity theorems for C^*-dynamical systems, 2006

849 **Katsuhiko Kuribayashi, Mamoru Mimura, and Tetsu Nishimoto,** Twisted tensor products related to the cohomology of the classifying spaces of loop groups, 2006

848 **Bob Oliver,** Equivalences of classifying spaces completed at the prime two, 2006

847 **Eric T. Sawyer and Richard L. Wheeden,** Hölder continuity of weak solutions to subelliptic equations with rough coefficients, 2006

846 **Victor Beresnevich, Detta Dickinson, and Sanju Velani,** Measure theoretic laws for lim–sup sets, 2006

845 **Ehud Friedgut, Vojtech Rödl, Andrzej Ruciński, and Prasad V. Tetali,** A Sharp threshold for random graphs with a monochromatic triangle in every edge coloring, 2006

844 **Amadeu Delshams, Rafael de la Llave, and Tere M. Seara,** A geometric mechanism for diffusion in Hamiltonian systems overcoming the large gap problem: Heuristics and rigorous verification on a model, 2006

843 **Denis V. Osin,** Relatively hyperbolic groups: Intrinsic geometry, algebraic properties, and algorithmic problems, 2006

842 **David P. Blecher and Vrej Zarikian,** The calculus of one-sided M-ideals and multipliers in operator spaces, 2006

841 **Enrique Artal Bartolo, Pierrette Cassou-Noguès, Ignacio Luengo, and Alejandro Melle Hernández,** Quasi-ordinary power series and their zeta functions, 2005

840 **Sławomir Kołodziej,** The complex Monge-Ampère equation and pluripotential theory, 2005

TITLES IN THIS SERIES

839 **Mihai Ciucu,** A random tiling model for two dimensional electrostatics, 2005
838 **V. Jurdjevic,** Integrable Hamiltonian systems on complex Lie groups, 2005
837 **Joseph A. Ball and Victor Vinnikov,** Lax-Phillips scattering and conservative linear systems: A Cuntz-algebra multidimensional setting, 2005
836 **H. G. Dales and A. T.-M. Lau,** The second duals of Beurling algbras, 2005
835 **Kiyoshi Igusa,** Higher complex torsion and the framing principle, 2005
834 **Keníchi Ohshika,** Kleinian groups which are limits of geometrically finite groups, 2005
833 **Greg Hjorth and Alexander S. Kechris,** Rigidity theorems for actions of product groups and countable Borel equivalence relations, 2005
832 **Lee Klingler and Lawrence S. Levy,** Representation type of commutative Noetherian rings III: Global wildness and tameness, 2005
831 **K. R. Goodearl and F. Wehrung,** The complete dimension theory of partially ordered systems with equivalence and orthogonality, 2005
830 **Jason Fulman, Peter M. Neumann, and Cheryl E. Praeger,** A generating function approach to the enumeration of matrices in classical groups over finite fields, 2005
829 **S. G. Bobkov and B. Zegarlinski,** Entropy bounds and isoperimetry, 2005
828 **Joel Berman and Paweł M. Idziak,** Generative complexity in algebra, 2005
827 **Trevor A. Welsh,** Fermionic expressions for minimal model Virasoro characters, 2005
826 **Guy Métivier and Kevin Zumbrun,** Large viscous boundary layers for noncharacteristic nonlinear hyperbolic problems, 2005
825 **Yaozhong Hu,** Integral transformations and anticipative calculus for fractional Brownian motions, 2005
824 **Luen-Chau Li and Serge Parmentier,** On dynamical Poisson groupoids I, 2005
823 **Claus Mokler,** An analogue of a reductive algebraic monoid whose unit group is a Kac-Moody group, 2005
822 **Stefano Pigola, Marco Rigoli, and Alberto G. Setti,** Maximum principles on Riemannian manifolds and applications, 2005
821 **Nicole Bopp and Hubert Rubenthaler,** Local zeta functions attached to the minimal spherical series for a class of symmetric spaces, 2005
820 **Vadim A. Kaimanovich and Mikhail Lyubich,** Conformal and harmonic measures on laminations associated with rational maps, 2005
819 **F. Andreatta and E. Z. Goren,** Hilbert modular forms: Mod p and p-adic aspects, 2005
818 **Tom De Medts,** An algebraic structure for Moufang quadrangles, 2005
817 **Javier Fernández de Bobadilla,** Moduli spaces of polynomials in two variables, 2005
816 **Francis Clarke,** Necessary conditions in dynamic optimization, 2005
815 **Martin Bendersky and Donald M. Davis,** V_1-periodic homotopy groups of $SO(n)$, 2004
814 **Johannes Huebschmann,** Kähler spaces, nilpotent orbits, and singular reduction, 2004
813 **Jeff Groah and Blake Temple,** Shock-wave solutions of the Einstein equations with perfect fluid sources: Existence and consistency by a locally inertial Glimm scheme, 2004
812 **Richard D. Canary and Darryl McCullough,** Homotopy equivalences of 3-manifolds and deformation theory of Kleinian groups, 2004
811 **Ottmar Loos and Erhard Neher,** Locally finite root systems, 2004
810 **W. N. Everitt and L. Markus,** Infinite dimensional complex symplectic spaces, 2004

For a complete list of titles in this series, visit the
AMS Bookstore at **www.ams.org/bookstore/**.